LIFEOVERTIME

Everyday Human Survival

Kenneth Jeffery Farnol

ISBN-13: 978-1508763420
ISBN-10: 1508763429

☥

SURVIVAL = LIFE / TIME

$$E = mc^2$$

The Author

Kenneth Jeffery Farnol was born in Scotland in 1942. A devoted family man he is now happily retired in Wales. Following uncertain beginnings and a disrupted education Ken settled down and went on to have a very rewarding and fulfilling life. He served in the Merchant Navy before becoming a Non-Destructive Testing Engineer in the Steel, African Mining and Aerospace Industries. Ken remains a restless free-spirit who loves the sea, mountains and world in general. He enjoys life, art, music and literature. He is largely self-taught, open-minded and still questions everything. He believes you never stop learning and is a member of the splendid University of the Third Age (U3A). Ken remains an avid freethinker, humanist and visionary.

CONTENTS

Prologue

Humans have survived for a very long time
Everything we need to sustain life is available
We have intelligence, energy and determination
We are aggressive, co-operative and controlling
We are an enigmatic and remarkable species
Shall we take a closer look at ourselves?

INTRODUCTION

There is no wealth but life.   *John Ruskin*

How often do we think about our day to day survival? You know - our everyday subconscious, unspoken, unthinking urge to continue life on earth - to be human, keep human and remain human for all time? We take it for granted. We just 'are'. Why do we not see it for the miracle it really is? Life is for living. It is as natural as breathing. Our physical survival of course depends on breathing, eating and sleeping. Abundant natural resources sustain our lives. Water too is essential to life. Yet there must surely be more to survival than mere physical needs?

We have many psychological, sociological and emotional needs too. Humans are thinking, curious and creative beings. Intelligence, acumen and insight make us what we are. We are a very advanced species. Our stunning imaginations and intellectual abilities enable us to see further than where the next meal is coming from. We can even ask ourselves metaphysical questions, such as: what is life about...and what is it like? We have more than just merely survived - we have progressed from the most humble origins to travellers in space. So why do we pay so little heed to our basic day to day survival?

Survival = Life / Time

The purpose and meaning of life is unknown but is an astonishing gift to be preserved at all costs. All species strive with, for and against each other to perpetuate their own place in land, sea or air. Human beings are no exception to this rule. Yet as intelligent beings we give surprisingly little thought to this idea. Survival is such an inborn and subconscious instinct that we take it very much for granted. We just get on with life. Yet we survive to live and live to survive on a day to day basis.

As human beings we share life with millions of other species. There are myriad resources available for life on earth. We are surrounded by organic and inorganic matter of infinite quantity and complexity.

Einstein's Laws of Physics state that Matter/Energy can neither be created nor destroyed. So all we can do is adapt matter and energy to ensure everyday continuation of life. The inexplicable kinetic energy of life derives from matter of course. All that energy, all that amazing life: food for thought.

Every substance, mineral and material needed for tools, shelter, food and warmth are available, if only we learn how to adapt them for practical purposes. Electro-magnetism, fire, wind and weather must have seemed like magic to our forebears. No wonder they worshipped the elements. No wonder they fell down and worshipped the Creator, the universe or the great Unknown if you prefer. They only needed to look up in wonder at it all; as we still do.

Man has always wondered. Yet in spite of our mysticism and spirituality we often remain fundamentally greedy, predatory and war-like. Whilst by no means perfect: at our best we are enlightened, ethical and caring. We are a conservative yet enquiring species. Our brains have taken us to the top of the food chain. Humans now control much of the world. Many question if we can continue to plunder seemingly inexhaustible resources. According to Darwin all species must adapt in order to survive. Humankind is nothing if not adaptable.

Many still worship their various deities. Does this include due reverence for Life over Time i.e. Survival? It is a question of mind over matter. We are all part of Nature. Regardless of theology we are all subject to the remorseless instinctive survival impulse. Many take it for granted: we just are - who cares? Yet everything we <u>are</u> relates to Survival. Everything we <u>do</u> is related to Survival. Everything we <u>think</u> relates to Survival. It is in Mind, Body and Soul.

Like human Faith: survival is all about hope for a better world. We are a curious race in both meanings of the word. Many of us seek to learn more about ourselves and the world around us. Therefore we must free our intellects from centuries of dogma. We need to open our minds and think for ourselves. Whether we realise it or not we are all born survivors. Perhaps we should take an unbiased and critical look at some ideas about human survival. We owe it to ourselves, our predecessors and our successors. It's in the genes.

Most of us simply get on with our lives as best we can. Yet there must surely be more to human survival? Perhaps we have compulsive and inescapable instincts to procreate and survive. Humans are no exception to this fundamental law of nature. Do the inherited genes of our myriad ancestors demand that they too did not live in vain? Why did we evolve in the way we did? So, is it true that everything we do relates in one way or another to our ultimate survival? For example we can be very competitive, controlling and aggressive when needed.

We successfully spread across the globe over many millennia. This required brains, courage and resourcefulness. We even had to fight each other from time to time for territory in order to survive. Nomads and warriors eventually became peaceful farmers, traders and citizens. We fulfil our basic needs in many ways. Humans are a truly advanced, remarkable and unique species. Our infants instinctively get up and walk. They recreate our primeval origins before our very eyes: onwards and upwards. What an interesting thought.

Humanoid beings have been surviving for a very long time now. As hunter killers we did not achieve this by altogether peaceful means. This may explain some of our less gentle traits to this day. Yet, perversely we co-operate with each other in order to ensure group survival. We use tools, communicate and look after one another. We control the animals so vital to our survival. We are spiritual, emotional and creative beings. We teach, learn and have ethical values. Many of us now live together in vast cities. Some of us are even quite civilised.

Our society is based on status and class. Some have money, comfort and pleasure. Others have nothing. So what has that to do with basic survival? Well, regardless of quality of life, we still remain survivors of one kind or another. The more we think about human survival the more miraculous it all becomes. Life is an ongoing and phenomenal gift to be savoured, enjoyed and appreciated. Life is for living. We are sensitive and sociable. We have minds, souls, memory and aspirations. We fall in love. We have shelter, families and friends. We even enjoy a good joke.

We are inventive, curious, amusing and have vivid imaginations. This goes without saying. We are a sophisticated social species working together in our common interests. Many humans still have faith and live in hope of better things. Some think for themselves; others think for us. We care and share; yet fiercely compete with one another. In fact there is a selfish, malevolent and cruel side to us which is very difficult to like.

Our leaders are vital but when they get it wrong then heaven help us. Our hereditary class systems are necessary but often unjust, cruel and divisive. We rely on law and order to survive; yet when society becomes too strictly controlled, multi-layered and *over-structured* then mass warfare and revolution become inevitable. Human beings are one of nature's greatest enigmas. Conflict and co-operation, war and peace, good and bad, love and hate; control, politics and status: these are all part of living. Are all these really related to the long term continuity of human life?

This life-study concentrates on day to day human survival or self preservation. From the outset we shall have to adopt a very unbiased, or 'metaphysical', approach. This simply seeks truths about life and being. This is nothing new. Good old Aristotle had it worked out long ago. The truth is we all live and survive. We give very little thought to this simple idea and take life for granted. We just 'are'. Yet we neglect many fascinating and unspoken topics of huge significance to our daily lives and the way we look at the world around us. In fact we may never be quite the same again once we get on these exciting new wavelengths.

This is neither a religious nor scientific treatise; although these important human survival philosophies will make their inevitable appearances in due course. In fact this project is metaphysical in the sense of asking fresh questions about the meaning of life rather than relying on outdated myths, arcane legends or complex formulae. Let's keep it simple. Well as simple as we can.

We shall perceive some disparities between the practical meaning of life: which is simply to be maintained at all costs and the less tangible purpose of life: which we may never know. This study is intended to be an unbiased, rational and logical analysis of life, time, cosmic energy and the mysterious Life-force itself insofar as they relate to our survival on earth. However, the origins and intentions of these ambiguous powers will probably remain as mysterious, mesmerizing and unknown as ever.

NOTE:
The concept of a Creator unavoidably appears in this discourse on survival. Whilst respectful of all faiths and deities; for our immediate purposes we must simplistically construe mankind's two major interpretations of the Great Unknown as follows:-

INNER - *There is the man-centred psychological and emotional concept of 'God' (Mind).*

OUTER - *There is the perceived Universal Creator of all Cosmic Energy and Being (Matter).*

For further discussion on fundamental religious beliefs, relative to human survival, attention is drawn to the Appendix: Defining 'God'.

This is a largely non-technical and non-partisan view of a neglected but fascinating topic of everyday relevance to each and every one of us. I wonder how my radical theories of an over-controlled and top-heavy status obsessed society will be received. What is ancestral memory or instinct for that matter? What will we make of human love and caring on the one hand and vicious human conflict on the other?

There will be a lot more for us to think about during the course of these essays and an open mind will be an essential attribute. I only hope my own mind remains as open as I might like it to be. We are embarking on a new and exciting way of looking at our very lives and being and what could be more important than that? After all we are a very inquisitive species. We are all human beings.

We shall encompass: life, evolution, caring, family, adaptability, tools, social status, control, and conflict. Additionally, life-science, religion and the intriguing world of metaphysics will also be discussed. So far as practicable I want to discuss rather than dictate my opinions on these sometimes controversial matters. These absorbing subjects are far less off-putting than you may suppose. We shall also look at natural selection, ancestral memory, work, needs and so on; using simple language wherever possible.

On the face of it seemingly complex ideas of physics, metaphysics, existentialism and telepathy may appear a little daunting. This is far from the case. It would be a great pity to miss these fascinating themes so critical to everyday human life. These can be just as simple or as challenging as you want to make them. Common sense, instinctive human curiosity and imagination are all you need. This foray into the relatively little explored field of human survival is broad and wide-ranging There are very many questions but not quite so many answers.

I have made some hesitant conclusions but you may well favour your own ideas about these matters. It is the discussion itself which matters.

This series of essays is intended to inspire everyday interest in these little explored subjects. It is hoped these survival topics will inspire both imagination and enthusiasm. Nowhere does the subject matter become over-technical or over-challenging but some topics may be slightly easier reading than others.

The project is designed more as a series of separate essays than as a continuous narrative. The free-ranging and free-thinking theoretical nature of this venture is aimed at both the general reader and the more esoteric market. It is again recommended that we approach this surprisingly unexplored subject with an entirely open-mind. This will be a journey of discovery consistent with that miracle of all miracles: we survive in our minds, our instincts and our very souls. What could be more captivating? The priceless value of human imagination will remain a major theme throughout. We shall think for ourselves, free from any misconceptions; well so far as humanly possible that is.

This discourse investigates the hidden wonders of the day to day survival of the human race. We shall concentrate on everyday human survival rather than that of the millions of other species with which we share a seething planet. This is about life in all its splendour. In passing we may marvel at the miraculous quantity of animal, plant and mineral resources available to us in order to enhance our human survival.

We are nothing if not an adaptable species in accordance with the theories of Charles Darwin (1809-1882). Survival is about life. Life is about survival. We continue to survive and gladly live out our allotted span on earth, yet very seldom give it much thought. Why? Perhaps we are reluctant to explore that which is largely subconscious, instinctive and natural. Many of us simply do not question anything which is so basic, given and fundamental as life. We are in denial of our own mortality. Pity really; as we are missing a great opportunity to try and understand ourselves a little better. This has little to do with the cessation of life. This is more about living our lives.

This presentation encompasses some of the less obvious and more abstruse survival issues which go largely unnoticed on a daily basis. My objective is to introduce the reader to the less perceptible factors at the foundations of life and human survival. If nothing else these should give us an insight into a fascinating and little researched way of looking at our very being. What could be more remarkable? What could be more appealing to our astonishing human imaginations? Our imaginations have got to rule supreme.

So to sum up: this presentation is quite general, discursive, yet intuitive, rather than specific by nature. It should simply be seen as a broad overview on the subject of day to day human survival. There will also be some straightforward further discussion on the practicalities of human survival such as the roles of tools, territory and farming methods. The basic conflicts between science and religion will also be examined at some length. There will be many more aspects of life and human survival of varying complexity to discuss. Yet we can only show the tip of the iceberg of this fascinating and little known subject.

Maybe we shall start to see ourselves in a new and fascinating light? If nothing else our inborn curiosity about ourselves may be stimulated by the following account. We may even gain a little insight into the intriguing world of metaphysics which is simply the search for truth about life.

So let us have a good long look at ourselves and don't blame me if we don't always see eye to eye. It is all reasoned speculation and we shall hopefully survive to tell the tale. Whilst some tentative conclusions may be reached there will undoubtedly be very many more open-ended questions for us to consider.

NOTE:
I have not discussed the practical and specialist life-preserving disciplines of Hygiene, Diet, Medicine, Welfare, Health and Safety and so on. These have significantly improved life-spans in the United Kingdom from 30+ in 1313, to 50+ in 1913 and 80+ in 2013. We primarily owe our present day low infant-mortality rates and high life-expectancy rates to modern public-health facilities in general. We owe a huge debt of gratitude to all those dedicated professionals who do so much to prolong human survival. This should go without saying in a project of this nature.

8

1   LIFE

The ultimate value of Life depends upon awareness and the power of contemplation rather than upon mere survival.

Aristotle

So what is life? This is about ongoing life sweet life - life as eternal motion - life as solar energy - Life-force generated by the sun, moon and stars? Is it perhaps electro-magnetic and something to do with the nerves? Is life bio-chemical? Life is literally fed by the food we eat. We are what we eat. Our endless search for food to sustain life is shared by all living creatures and there it is all around us. The oxygen we breathe must help - life as a force for good - life as the ultimate challenge - life to be lived to the hilt - experience, fulfilment and even happiness. Life shared with others.

Life need not be a lonely place. Life can be what we make of it. Life gladly handed on - life, loving and living - life growing older and maybe wiser. Life and learning - learning about ourselves and the world we live in. What could be better - life as the greatest miracle of all time? - Life with all its ups and downs - living, creating and imagining. Life to be savoured, enjoyed and maybe even understood - well we can try. Yet we must all live and die. The Epicurean philosophy of: 'eat drink and be merry (for tomorrow we die)' appears in the Bible of all places. *Ecclesiastes 8:15, and Isaiah 22:13.*

Humankind is unique in the animal kingdom in having sufficient intellect, imagination and spirituality to even think about the meaning of life. Yet all are driven by that other great enigma - instinct. It is hardly surprising that the mystery of life lies at the heart of all human faiths. No wonder practical survival is so understated when compared with religious and spiritual aspects of this challenging intellectual concept. It is an inward and individual Life-force of universal magnitude. Life is the true enigma. No one knows its true purpose.

Was life by accident or design? It is in our subconscious and is a secret obsession: yet, oddly, we hardly dare to think about it. Is the awesomeness of life too much to take in? Do life and fear have some strange affinity with each other? The reality of life is positive and the emotion of fear is negative yet they are both mutually symbiotic. You simply cannot have one without the other. Innate caution and fear of the potential loss of life even keeps us alive. Let us look a little further into these amazing ideas.

Life is an enigma - the first great mystery - the last great question - the great Unknown. Is it because we don't really understand life that we don't really question it? Is this why we give life so little thought and take it so much for granted? We simply do not and cannot understand something that is as natural but intangible, ethereal and inexplicable as life. It is a gift beyond price. What is the mysterious Life-force? Life over Time is the best thing we have got. Yet it is fraught with difficulties.

Survival is the greatest of gambles for many. In general we seek to survive at all costs. Yet for some less fortunate than others life can be compromised by unforeseen events. Life truly is a lottery.

Periodically some of us are even subjected to all the unwitting but inevitable horrors and sacrifices of war. What does war do for survival? How can this happen? This is the most difficult thing to come to terms with. There is perhaps some justification for defending against aggressors. This is probably a natural and primitive survival mechanism - but global war across continents?

Some aspects of our unfortunate propensity to make war upon each other will inevitably be looked at in due course. Attempts will be made to recognize these for the major assaults on human well-being which they are.

Are we in fact born control freaks that try and control ourselves, the world and everything around us?

This project is more about understanding ourselves for both good and evil too. This project is also to do with rich, vibrant and continuing life. It is about love, marriage, procreation and caring in society. It is about our positive achievements; tools, spirituality and civilization. It is about many things. How many minerals are present in our bodies? Are we really part made of water? What is our primordial connection to it?

Why did we migrate across the oceans? Was it hunger? Or were we hungry for adventure? Or was territory what we were really after? What of our biological place in the great scheme of things? We understand microbes as the building blocks of life on earth. Similarly In the scale of things we are mere micro-organisms in terms of the universe. There are worlds within worlds and so on. We maybe think we are gods but in the vast scheme of things we are quite insignificant. I for one gave up these futile questions leading to madness long ago. My ego was not up to the job. No wonder we are such an ambiguous species.

To bring in notions of metaphysics: we can still ask ourselves a few challenging questions as follows. Are we an accident of nature or are we in fact put here to some ulterior purpose? Have you got any answers? What is energy? What is Life-force? What is life? What is time? What is survival? So let us look at the formula again:-

$$SURVIVAL = LIFE \, / \, TIME$$

Do our gods lie somewhere deep in this simple equation? I know 'God' turns up quite often but is bound to in the nature of this investigation into our very being. Time will be discussed before too long - all in good time. Good Old Father Time - do we really comprehend just what time is? It certainly is the dynamic element of energy, life and survival.

For some closer understanding of both life and time we now need to consider the ecology of existence. For the purposes of our argument let us look at the very foundations of biology. Let us look down the microscope and examine the very constituent molecules of life. Never mind the great unknown macro Gods of the universe for now. Never mind the pantheistic Gods of Nature. Never mind the human Gods of the mind: or whatever vague images we conjure up from deep in the subconscious or ancestral memory. Never mind all the other humanlike Gods of so many different types.

No matter how fervent our faith there is no escaping the Micro-world at the root of all existence. What of the Micro Gods: the atoms, the molecules, the DNA, the building blocks - all the biology, elements, synergy, maths and science of existence - the intimate, minute and intricate detail of it all? You may never be the same again once you have looked down that imaginary microscope. What of the quantum physics and microbiology of the dynamic matter/energy in motion which is life?

What is this 'kinetic energy', or Life-force (Chi), in each and every one of us? What is real and what is in the mind? Is it all some kind of mystic or mythical challenge thrown down to perplex and mystify us?

Again what is time for example? Is time the mathematical yet inexplicable dynamic constant we think it to be? How do we explain time - that great enemy - that mysterious entity by which all life is ruled? Is Time the real God or merely part of a greater purpose?

Time and tide wait for no man (Burns).

Or is energy (tide) the ruling force for which time is the dynamic factor?

Evolution over time is of course a well trodden route but the concept of survival as an academic topic in its own right is far less well known. Evolution, as such, can be researched from layers of evidence. Survival is far less tangible. It is mainly a matter for idle conjecture if we ever even bother to think about it. It is grossly overshadowed by the theological studies so beloved of humankind. It is too everyday and pragmatic to have much mystic glamour associated with it. Pity really. No wonder there are so few readable books on the subject but this is not to deny its relevance to human life.

Survival must hang by a thread for some but be of major significance to the life chances of others. We know deep down that there is only a certain amount of wealth and prosperity to go round. We know the inequalities of life. We all know the bitter truths behind the much quoted parable: *the feeding of the five thousand.*

The inborn impulse to survive, now and into the future, is subliminal, inescapable, and utterly compulsive. If we are fortunate enough to have children the same rules apply. It is all about the long term continuity of our species; unspoken, unappreciated but critical. Survival to the theoretical end of time; (which according to Einstein will never happen as matter cannot be created or destroyed) would presumably see our society changing, evolving and developing, just as it has in the past, by adapting to survive.

The primary theories of human survival, as a potentially broad academic discipline, are believed to be grossly unexplored. It is hoped this layman's survey of this remarkable subject will stimulate further speculation and perhaps even some conclusions of some value to humankind. What could be more relevant to the study of the humanities in general? Surviving in all we do. Whoever we are, wherever we are, whatever our faith, what could be more fulfilling than life itself? There are of course many sides to the wide-ranging speculations in these essays.

Much of which follows is constructive supposition (guesswork) yet it is trusted that a certain logic prevails throughout. Above all it is hoped that these essays are more thought-provoking and entertaining than conclusive.

The survival of the human race is perhaps our first unspoken priority. Why else do we live? We live to survive and survive to live. We are ruled both by nature and nurture. What of procreation, love, and caring? Such is the importance of the family at the heart of society. In spite of the infallibility of these arguments; why is day to day human survival so little written about?

This is a vital, compelling and compulsive subject by any standards. So if it is so important - why do we take our survival so much for granted? Are we reluctant to face up to our ultimate mortality? Is this why we are so loath to ascribe the realities of daily life to such a tenuous subject? Other than as an intellectual exercise: what does the theoretical study of survival do to promulgate actual practical survival? Probably not much - but what a fascinating subject it really is - if we can only open up our minds to give it some serious thought.

Our natural curiosity and imaginations will then do the rest: something not to be missed.

Survival, like religion, is very much about hope for the future. Yet, if we are honest with ourselves we are not quite as agreeable a species as we might like to think we are. We are often greedy, selfish and even downright nasty. The enigma of mass human conflict is particularly harrowing, puzzling and challenging. Yet the courage and gallantry of those who fight for, with and against their fellow human beings is inescapable. Yes, we can be both noble and pitiless at the same time. We may not always like some of our less savoury human traits but must always believe that we can hope for a better place.

The conflict between good and evil has always formed part of the human dichotomy. Whilst this is not in any way a religious treatise: it is strongly believed that faith in a universal deity is a very natural human instinct. We are nothing if not spiritual. We are alive and this is the greatest gift we possess - the gift of life - what could be more worthy of handing on to our progeny?

The sad truth as a competitive, whilst not entirely unsympathetic species, we know deep down that it is 'us or them' when it comes to enjoying the fruits of the earth. Otherwise let us never forget there are millions in the so called third world living hand to mouth at barely subsistence levels whilst we in the west suffer from chronic over-indulgence, greed and selfishness.

These poor third world cousins are the true experts on survival at its most basic level as are the downtrodden and dispossessed in our affluent western cities (that we see but don't see at the same time). Whilst there are always people living tragic lives of poverty, pain and desperation: many of us nowadays survive at an altogether more comfortable level.

Survival and humanity must go hand in hand. What could be more fundamental to our treatise on human survival than a long and hopefully interesting chapter about human evolution from humble beginnings to an age of communications, space exploration and computers.

2   EVOLUTION

It is not the strongest of the species that survives, nor the most intelligent that survives. It is the one that is most adaptable to change.

Charles Darwin

Just imagine a huge tree of life showing all the living species you can think of. I mean every animal, insect, crustacean, worm, seaweed and plant there is. Now think of an upright human-being wearing a long cloak and waving a big stick standing at the very top of the mammal branch. Could the human species have survived without a stick as the first tool? Could humanity have survived without the other branches of life? Without tree branches to grasp and to climb: then where would our first tools (sticks) have come from?

The plants and trees are of crucial importance to humankind and many other species too. Many of us are parasites who live off each other. The food-chains include both animal and organic vegetable matter. As a general rule: the animals have brains and move about the earth. The plants, which many animals eat of course, have no brains.

Plants stay where they are rooted in the earth or sea (although they too spread far and wide). If you think about it: this is quite a good formula for survival. Without brains, plants and each other we starve. Yet we all share the same air, water, soil, vitamins and minerals. Miraculous! Now tell me that life is boring. So don't feel too bad about eating that chicken or even that cabbage: it keeps you alive.

Never mind the vast quantities of natural mineral resources available to humankind for now. Without innumerable fellow creatures and vegetation to catch, eat and grow; then how would we have developed and survived as we still do? It is amazing to think that all living animals, humans and plants are descended from humble common origins - and over such infinite time periods too. Yet we have already seen that Survival = Life over Time: a lot of life over a lot of time.

It is all about Darwin's adaptability of course. In some remote epoch we somehow ended up sacrificing two front legs to use as arms and stand tall amongst our fellow quadrupeds. How did that happen? There are other strange and ungainly animals too, like kangaroos, trying to emulate this but with varying degrees of success. Some (unconsciously) went for wings to fly with and used their beaks as tools instead. Others developed fins or web feet to swim with. Others did both. Yet we humans went that much further with our opposable thumbs and skilful fingers. We made tools. We made the wheel. We can swim too.

We needed hand tools so that we; by team effort, could eventually design and make aircraft to fly (and ships to sail) across the oceans: so how remarkable is that? We more than just adapted that is for sure.

Why do we never stop to think about these things? What is wrong with questioning the nature of our very being? This is why we got where we are today. Our already miraculous lives can surely only be enhanced by this very natural curiosity to understand just who we are. Why were we given our huge brains (stuck right on top of our odd upright bodies) if we don't even think to use them to question our basic origins? We should just occasionally think about these things if only to enhance our knowledge of life itself. Believe me, nothing could be more fascinating.

Perhaps the following flippant remarks should be reconsidered in the light of our new way of looking at ourselves:-

Take a good look at yourself!

Who do you think you are?

What are you like?

Seriously, 'take a look at yourself' next time you look in a mirror and wonder. Who am I? You may never be the same again. You carry many centuries. What could be more crucial to all that we are - and may be - in the future? Is it too much to take in? Is it all one great big miracle? Or is it a great big joke? Yet here you are - but survival is no joke is it? It happens. Think on!

The awe inspiring biodiversity, creation and survival of all these mutually interdependent life forms on sea or land are breathtaking, overwhelming and incomprehensible. This is surely above and beyond any concept of God we might hold. This is perhaps why God was created in our own upright image - to bring us down to earth. The infinity of mass, energy, space and time of the universe must surely be the real origins of all this living matter.

The very atoms, electro-magnetic forces, kinetic energy, proteins, DNA, stem-cells, chemistry and physics of our being must be the building blocks of our lives on earth. Good heavens, there are 7 billion + of us in just our own species. There are multi-billions of other beings, bacteria and organisms to which we owe our very existence. Well we must all eat other living things to survive. Our very existence is indeed some kind of miracle. Life is sweet and we live in hope.

We are very puny members of an infinite universe. It is like a vast cosmic jigsaw. When we look at our imagined tree of life the evolutionary time-scale of these interlocking and sometimes symbiotic organisms is again cause for wonder. Most living matter is constantly competing, co-operating, changing and striving for dominance over the others in order to survive.

Other less ambitious but eminently successful creatures happily remained simple, inconspicuous and unchanging over millions of years. The sheer quantity and diversity of living matter is so phenomenal as to beggar belief or understanding. I am told there are something like 8.75 million known species on earth - all surviving one way or the other. Some have changed more than others. There have been mass extinctions, such as non-avian dinosaurs. Did they get too big for their boots? Yet the more adaptable ex-dinosaur flying birds still live-on.

According to some sources Earth formed some 4.54 billion years ago. Life appeared on its surface within the order of 1.0 billion years. Primates, from whom we are descended, are a very recent animal. They emerged something like 60 million years ago. Since then there have been a variety of ape, humanoid and monkey species. Many branches of these have now become extinct. Archaic Homo-Sapiens, the forerunner of anatomically modern humans, evolved between 400,000 and 250,000 years ago.

It is believed that advanced humans originated about 200,000 years ago in the Middle Palaeolithic period in Southern Africa. Is this the age of our human God? If so, He was a bit late for the Creation by many millions of years. Of course we now have information about our origins not available to the less well informed biblical authors of Genesis.

Nothing changes really; the more you research our modern knowledge of the origins of man the more confusing it all becomes. Millions of years are bandied around but one thing you can be sure of is that we have been around, in one form or another, for a very long time indeed. Yet our progress has been nothing short of remarkable.

At various, not yet determined periods, early humans may have migrated out of Africa into Asia and Europe and began the long process of colonizing the entire planet. 'Modern' humans evolved something like 50,000 years ago. They are believed to have reached the Americas, by one means or another, merely 14,500 years ago. One wonders if the famous nomadic semi land-bridged Siberia to Alaska route was the only one - as seafaring across the islands of the Pacific and along the coasts of North, Central and South America at various other periods was also feasible. How else could you get your green card?

We are still very much developing as a species. One estimate puts the ability of the Earth to continue to support future life at 500 million years. How successful we shall become remains to be seen. Some of our fellow carnivores go back quite a long way too. Crocodiles, for example, go back to the age of the dinosaurs: 240 million years ago. Sharks go back even further (420 million years). Some of these opportunists became so large and efficient that there was little need for any significant further development.

Yet they have many variations according to their geographical locations, climate and food types. The very largest of the sharks eat plankton and are totally harmless. The next largest in size are fearsome predators. It is all about adaptability. Thus rules of versatility still obtain. It's almost as if there is a niche for every conceivable type of life that can find the means to survive.

We have many carnivorous competitors. Does their position of strength at the top of their own food chains challenge Darwin's theories about change? Yet they obviously changed at some point in their evolution. Just look at them. Some of our nearest land based competitors such as the big cats, as we know them, seem to be of similar age to mankind. So, by and large, species not only survive but evolve and change as if driven by some unknown force. I am suggesting this is something simplistically called survival. Or a natural instinct for survival shared by all species.

It is interesting to note that humanoid fossils and artefacts from the past few million years are often to be found in the African Rift Valley in areas that alternated between lakes and dry savannah. Global warming and climate change is nothing new. These marginal areas must have been ideal hunting and fishing territories.

Humankind obviously used violence to survive. It is interesting to note that the major quarries of most land-based carnivores are the predominantly less aggressive herbivores. The hunters indeed have eyes at the front of their heads and the hunted have eyes on the sides of their heads. Were they designed and destined to be the hunted and us the hunters? Only the most alert and swiftest members of the defensive group may avoid capture until age or infirmity catches up on them. The presence of often vicious horns may sometimes help in a defensive role but these are more likely to be status symbols of superiority and competition, within the hierarchy of the herd; to enable ultimate survival of the majority.

Many African animals die a pitiful death in the bush. Homo-Sapiens was often able to avoid this type of violent death by using varied hunting habitats within reach of defensive shelters where fires could be lit and communities be kept in relative safety. Flexibility, curiosity, imagination and adaptability were amongst our primary survival tools.

By eating herbivores we, as hunters, were spared the necessity of grazing directly for many hours on vast amounts of grass. Although we learned that much could be done with the quern-ground seeds of some types of grass when harvested and made into bread of one sort or another.

Yet we too may survive on various types of plants; so are the most versatile of creatures. In fact we tend to thrive on a mixed diet of meat, fish, seeds, vegetables and fruit. Our metabolisms must have evolved accordingly. I believe our versatility and ability to hunt and eat virtually everything that moves, or does not move, in all climates on land or sea must have enhanced our chances of survival. We are what we eat. Other than eating fish we seem instinctively reluctant to eat our fellow carnivores - but would do so if necessary.

We almost worship our feared fellow killers such as tigers, lions and leopards. They too are capable of extreme violence. There is something about their sly, sleek lines and lithe beauty which appeals to our deepest instincts as fellow killers. How we would like to be like them. Essentially we respect the intelligence, strength and cunning of our fellow carnivores. These are of course the ultimate survivors. We even make pets and hunting assistants of the more amenable ones; such as dogs.

Birds of prey are often larger, cleverer and swifter than their less voracious and tamer seed-eating cousins. Chimpanzees will occasionally hunt and eat monkeys but are mainly herbivores. All meat-eating hunters require imagination, intelligence, observation, stealth and cunning in order to outsmart their victims. All hunters also need the ability to elude their fellow predators or hostile and uncooperative prey when this became necessary. Fight or flight became the order of the day.

In my opinion it was our ability to capture, tame, and breed our own supply of animals for meat, milk, skins and transport etc. which gave us a significant edge in our ability to survive. In time we became nomadic, semi-nomadic and sedentary herdsmen, shepherds and horsemen. There are still people living as nomadic herdsmen in semi-desert areas of Africa and Asia in particular.

By these means and later by growing crops; we could create the necessary surplus for barter, trade and for civilization in general to expand. What a revolutionary transformation of primordial and primeval society that turned out to be. We are still adapting to this fundamental change from the long established tribal to the relatively recent urban social order. My theory: that the onward march of civilisation profoundly influenced us for both good and evil remains to be seen.

Before the relatively recent arrival of farming the cunning, tools and teamwork needed for survival as carnivores was of course essential on our long journey to the top of the tree of life. It is amazing really. This was no place for shrinking violets. World domination could only happen if we became top of the tree of life. This would have to be organised as a team effort to even stand a chance of success. I would suggest that any one of the above examples of carnivores would ultimately be capable of ruling the land by being the fittest. That is excluding the ones in the sea.

All species would of course need to nurture, feed and teach survival skills to their offspring. Territory is important to all species and humans are nothing if not territorial.

Most species will, of course, fiercely defend their family homes and territories and we humans are no exception. Too many people inhabiting one territory will logically result in attempts to expand these territories in order to survive. This of course often leads to bitter conflict with neighbouring populations in similar circumstances. Should these other territories belong to those speaking another language or of different cultural values: then maybe there will be further problems of suspicion, intolerance and general lack of mutual trust? This may occur between mutually antagonistic local tribal territories or even across international boundaries.

There were certainly some so called warrior tribes, such as our chosen race of Vikings; who eventually settled somewhere else, intermarried and became established farmers, craftsmen and peaceful citizens.

The role of human children in terms of survival is of particular interest and remains an interesting field for further study. Children were traditionally and often inhumanely used as virtual slave labour in less enlightened times. On the other hand their natural affinity with animals has been of great benefit to the survival of the species.

It is entirely feasible that some time after the first man fell out of the tree with his new found stick in his hand a very important apocryphal event happened as follows:-

One fine morning the children of the nomads came across some abandoned wild dog puppies to take home. These animals were so intelligent, cute and playful that no one would eat them. They turned out to be so loyal and friendly that they would even retrieve the toy sticks being thrown by the children; who they would have regarded as their surrogate parents. These dogs were then readily trained to hunt, bring back rabbits, kill vermin and round up large animals to the benefits of all concerned. They even gave warning of danger and became generally indispensible to the survival of mankind.

When the sticks were eventually made into fences; the dogs could help to drive herds of large animals such as deer into traps. These traps evolved into corrals with grazing where animals could be kept until needed for breeding, food, skins and antler horns. Their dung became an extra bonus as a fertiliser. Animal dung could even be mixed with other bonding fibres as a primitive building and insulating material.

Permanent buildings could be constructed in accordance with materials to hand. In lieu of stone; clay could be used to make bricks and many traditional types of houses may still be seen to this day. Simple tropical clay and pole structures or thatched roofed dwellings of one sort or another still survive in many places as they are quick and easy to repair or rebuild as required. Traditional slash and burn husbandry still exists in addition to more permanent farming practices around the world. Thus farming of both vegetation and animals led to more permanent settlements and acquisition of suitable territory.

Grass was of course a vital commodity for grazing animals. Some of the best grassland would be used for intensive grazing and poorer grazing over wider areas would be utilized in the manner of present day hill farming. Both animal and arable farming would be practiced according to the suitability of the land and irrigation available. The importance of territory for survival then became paramount.

There is a further trait concerning the role of children in our natural instinct for survival. To witness a baby of around one years of age standing up and taking their first steps is a realisation of the first tentative steps taken by mankind in order to eventually take control of our environment. This is part of a long childhood and learning process in order to survive in the way in which we have evolved.

Other less predatory defensive creatures such as horses, deer and antelopes learn to stand immediately after birth and attain full maturity by the time we are just beginning to talk to one another. Not only do human children, of say 8 to 12 years old, have a close natural affinity with animals of one kind or another but they have a predilection for primitive curiosity, creativity, exploration, den-building, forming gangs, playing 'house', hide and seek, teamwork, competitive games, mock battles and so on. The list is endless. Remember?

I have long believed that a basic understanding of the evolution of human-kind is symbolised right there by observing our children from first getting up to walk to the formal customs and rituals of adolescence. The conservatism of youth often comes as a surprise. Yet peer group pressure is a powerful and natural bonding mechanism. Much of this is social, natural learning and even high spirits but surely much of it must originate in our natural instinct for survival?

Dogs, as already pointed out, must have played a very significant role in the development of humanity. Man's best friend was followed by goats, sheep, cattle and horses and the rest is history. Then some piglets were brought home and some hens too. Thus bacon and eggs were invented to ensure not only basic survival but a state of complete luxury. The constant daily supply of essentially portable eggs and milk as high protein meat substitutes must have been of particular value. Skins, horns, bones, sinews, wool and butter from livestock were all important aids to human survival.

It is of interest that the Masai people of Kenya still maintain their cattle as highly valued stock but draw both milk and blood from them on the hoof; again for much prized protein. The principles of communal evolution in co-operation with animals should now be a clear indication of their essential contribution to our survival.

The evolution of arable farming around 10,000 years ago would again have led to the early civilizations of the Middle East in areas where water was available. Surrounding semi-desert and desert areas suitable mainly for goats and camels remained the territory of nomadic tribes such as the Bedouin to the present day. These too would know all there is to know about survival in a harsh environment.

The development of animal farming in a variety of terrains long preceded the evolution of arable farming where fixed water supplies and rich alluvial soils were essential for success. It is generally believed that arable farming of seed-bearing grasses capable of cross fertilization, such as 'emmer' wheat, first appeared in the Middle Eastern fertile crescent as recently as 10,000 years ago. In due course this early farming led to ancient market cities such as Byblos, Damascus and Jericho circa 5000 BC.

Further advanced civilizations such as ancient Egypt then developed in the region. Other types of seed crops, tubers and fruit such as dates could also be formally grown, harvested, processed, traded and distributed for animal and human consumption in due course. Together with meat and fish a varied and mixed diet of proteins and other health giving foods could be grown to supply a wide range of populations.

Essential commodities such as salt were also highly valued and traded over vast distances for those far from the sea living in semi-desert and savannah subsistence farming regions.

At the present time approximately one third of the population of the world work on the land. Prior to the Industrial Revolution, before around 1800, the vast majority of workers were farm workers of one sort or another. The land pervades our genes. The survival gene pervades our families.

We can now return to some of the underlying theories of both evolution and survival. Darwin's theory of evolution is all about the change in inherited characteristics of biological populations over successive generations. This living matter broadly includes species, genus, organisms, micro-organisms, molecules, proteins, amino acids and hereditary/genetic material such as DeoxyriboNuclear Acid (DNA).

Darwin would have been unaware of the finer points of genetics, DNA and the like. However, these have only served to enhance his theories in modern times.

There are further biological classifications in terms of plants, animals and so on. All these species and living matter in general also obey the rules of physics. These state that matter may neither be created nor destroyed but may be converted to kinetic energy such as the Life-force where protein is converted into physical strength, brain power and so on. This is the (literal) food for thought already mentioned.

Can we conceive how much energy has been expended on our very being - the quantities of food, gases, minerals, resources and recycling needed to sustain all that living energy? No wonder the sense of living heat/energy given off by say a human city or an ant colony or even a tropical rain forest are so palpable. Who knows how many species will evolve or become extinct long into the future? With up to about 9 million known living species we can do little but wonder at the purpose of it all.

Humans, as represented by Homo-Erectus, are perhaps only about 2 million years old and therefore mere twigs or saplings in the tree of life. Yet we have achieved control over not only our own lives but those of the many other living species on which we rely for food and shelter.

The long and successful survival of the human species is something we often take for granted. The fittest are of course those best adapted to survive - not just the physically strongest.

There are countless human survival adaptations. Can you think of any? How about: caring, competition, imagination and curiosity for starters? Curiosity is only one of many. Perhaps innate curiosity about ourselves and the world we all live in has been a significant adaptation in aid of human survival?

Brain power must surely count for human beings at least as much as physical strength. Is it all a question of human ego and sensibility? Have we got some cause to feel guilty about whom we really might be? Why do we often deflect attention away from too much self-analysis? The human animal is just as fascinating as all those other animals with which we seek to identify. How like us they behave - the great apes in particular of course.

Where do we all fit-in in relation to other types of life? In fact if we are honest with ourselves we quite often eat other forms of life. Do other species have a guilty-conscious about killing one another for example? This seems unlikely and maybe we humans look at things a little differently. We may be fellow predators but would like to think we have some finer feelings.

Evolution is all about both Natural Selection and Survival of the Fittest. What makes us the fittest i.e. best adapted or suited to survive? Where do we start? Perhaps we could start by looking at our ambivalent human characteristics of competition and co-operation.

Yet as already suggested how can human conflict be part of a survival strategy? What indeed of our inherent desire for wealth, status and control? Why must these always be at the expense of our fellow human beings? Why are some of us so selfish, greedy and hypocritical? Perhaps this is why our continuing survival is so little thought about? Am I treading in contentious territory here? Are we in denial of our inherently predatory natures?

There are many things we might not like about ourselves if we really thought about it. Are we hesitant to strip away the gloss of our religious and ethical veneer to acknowledge the primeval beast lurking beneath? How honest are we about ourselves? Surely we can't be all that duplicitous?

What have been the effects of our traditional moral and ethical values on civilisation and survival in general? Is 'good' preferable to 'bad' to ensure the continuance of the human species? What of our traditional religions? Should our ancient and revered faiths still be continued in the face of contradictory scientific discoveries?

Do we 'live-on' in heaven or in the genes of our descendants? Is religion all in the mind or is there a religious gene? Which is right, the Periodic-Table or Genesis? How can we reconcile these differences? Is there such a thing as ancestral memory? What about Charles Darwin and his theories of evolution, adaptability and survival? What of the practicalities of survival?

Why did we stand up and become nomadic hunters who roamed the wilderness for millions of years in order to survive? What of the use of tools? Was this the essence of our successful survival in a hostile world? With tools we could make fire. We could develop weapons. We could fashion clothing, boats and shelter. We could hunt, capture and tame animals vital for our survival. We could even learn to farm and grow crops and trade with each other. However, this is a comparatively recent development over the last 10,000 or so years: a mere drop in time.
Indeed we are an adaptable species.

Ultimately we learned to create cities, technology and even, amazingly, fly around the world. We are a virile, social and successful species who must have earned our place near the top of the tree of life.

We are at the pinnacle of something we blithely call civilization which has now, with a few setbacks, gone on for perhaps 10,000 years or so. This is all about evolution so how has civilization evolved?

If we agree with Charles Darwin our inherent adaptability is a major reason for our survival. It is suggested that our very vulnerability as a species forced us to become cunning, resilient and flexible. Fear has long driven us to survive by seeking warmth, security and shelter. It will be seen that when we were forced from the trees and adopted an upright posture and could grasp tools; then we could roam far and wide in search of our prey.

Our brains developed to such an extent that our whole physiological configuration significantly altered and developed in order to accommodate this primary feature.

With brain power came the power to imagine our gods. These were often in our own image. Our relatively complex means of giving birth, often needing assistance, together with difficult infant care and a long subsequent period of development and education changed our emphasis from that of mere survivors to one of enhanced intellectual and physical beings.

The huge significance of plants and animals to our success is very clear. This makes for an awe-inspiring proposition. This interlinking of species even extends, almost absurdly, to the fact that we can tame and train animals so that we can ride on their backs. No wonder that our ruling classes are so closely identified with horse riding. We even practice selective breeding of other species by controlling the reproduction of domestic animals.

We are a compulsively controlling species when all is said and done. Well some of us anyway. The writer admits to neither liking to control (or be controlled by) anyone but accepts that control in general is a very human characteristic. Here we approach one of our first inescapable conclusions. Like it or not we humans are a population of controllers.

It is acknowledged that: even in the face of hugely increasing world population and gross inequalities; survival prospects for some of us at least have never been better.

Questions of birth-control, disease-limitation, vaccination, diet, prolonged life-spans, medicine in general and the like: relative to survival; are highly significant of course. However, these are beyond the remit of the writer for the time being. We still have a great deal of work to do on this inexhaustible theme. True to life these ideas just keep on evolving. Evolve, adapt, think and we may just get there in the end. Maybe our survival was more than just pure chance.

As a recurring theme, literally everything we do both consciously and subconsciously is based on our own survival and unique evolution as a species and that of our offspring long into the future. This is the long path towards whatever destiny awaits the human race. Have we survived so well up to now that we don't even need to give it a second thought? Or do we? Do we really care? Maybe caring should be the title of my next chapter on survival.

3   CARING

Want of care does us more damage than want of knowledge

Benjamin Franklin

Do we really care? Surely we are all caring human beings aren't we? Well most of the time. When we say we don't care about anything: do we really mean it? What has caring to do with the survival of any species? Well certainly many of our warm blooded fellow creatures quite naturally and instinctively care for their off-spring.

Some types of fish even protect their young by sheltering them from danger in their mouths. Others have hundreds of abandoned offspring safe in the knowledge that maybe one or two will survive. However, as a general rule caring is vital for survival. We survive by nature and nurture. You could not have one without the other of course.

As humans we essentially care for the welfare of each other: for our families, communities and even for our planet. In spite of an insidiously dark side to our human natures, as discussed in the previous chapter: surely we instinctively care for human survival above all else? Not just present day survival but continuing existence long into the future.

Caring must be one of the best and most exceptional of our human attributes. Both men and women love and care: each in their own parallel ways. The intimate and tender loving care of the mother in general is supplemented by the complementary loving care of the father too. Surely instincts for caring and love for humanity in general are closely related?

Perhaps love lies at the heart of what is best in the human race. No matter how dispassionate one may be about the sometimes harsh realities of human survival it would be impossible not to be moved by some of the moral challenges encountered during a project of this nature. Survival, like religion, is often about living in hope of better things.

Quality of life is not the preserve of the exalted rich, beautiful and famous: it is about society in general. Some of us are indeed all too human. We must never forget the realities of human selfishness, hypocrisy, cynicism and disdain which often blight our so called moral values and those of many of our all too human institutions and political parties. We may even naively believe that these organisations act in the best interests of all of their members.

Many of our human institutes are founded on vast wealth and some are of course better than others. Surely a good society must care for all, not just some, of its members. Thus the concept of survival is of universal importance to us all whether rich or poor. Obviously our laws, basic morality, decency and self-discipline are vital to the survival of our species.

We may of course have divergent views of what is moral or amoral but common humanity to all must surely be the bedrock of civilisation? The warmth, caring and humanity of the extended family has traditionally suited some cultures. Perhaps extended human survival in terms of extended family life is the ultimate goal of this seemingly admirable concept.

We are all subject to Darwin's theory of change and adaptability. Different cultures have different survival problems in terms of climate and geographical origins. In fact the best survivors may well be those living in the most challenging of environments. The Inuit people for example, living in extremely cold surroundings, are especially noted for their courage, versatility, ingenuity and survival skills in general.

It is proposed that we must care for one another in order to survive as a general principle. I am very aware that there are low paid professional 'carers' and unpaid family carers for whom little enough assistance may be available. This does not always reflect well on our nominally caring society. I am also aware that there are fighting forces caring for our national homelands. This idea of caring has many facets. Caring for all ills must be across the spectrum, universal and available to all for a truly civilised society.

Whilst remaining strong, resourceful and adventurous as individuals surely the survival of our species is best served by caring for one another? We may even have to face up to the ironical idea that we sometimes must fight to show we care. We must inevitably be prepared to fight for justice or against tyrants and bullies in our midst should it become necessary. This is in our natures.

There are many strange ironies about caring for each other. Caring must be international and bodies such as the United Nations must become far more caring and less political for true impact on the welfare of the world in general. If only our caring institutions were more able to prevent abuses to the welfare of mankind before the event rather than after.

Caring, ironically, often requires the same discipline and organisation as human combat. For example we are all aware of moving images on television of fully trained fighting soldiers compassionately saving lives following natural disasters. If only the armies of the world were trained to save lives instead of destroying them. We must of course be very conscious of the great caring charities of the world. These often do the work that we might expect of our elected governments by proxy.

Some governments do in fact provide a great deal of social care and welfare to the needy in civilised communities - but it can never be enough - especially in prosperous countries with untold wealth at their disposal. The British National Health Service (NHS) for example must surely be a beacon of care - and an inspiration to all who look to a civilised society to care for them in their hour of need - pity that money can buy preferential fast track medical treatment even within the publicly financed NHS.

However, regardless of principle, who would not do anything necessary for personal survival if they could afford it? Again this underlies the remorseless human competition to survive (literally) at all costs. Survival and the caring that goes with it are manifest in many forms.

The writer finds these harsh realities of the survival conundrum in general both perplexing and distressing. The true reality of life is so often at odds with us so-called liberals (myself included) who go on and on about a caring society. We must never lose sight of the cruel realities of life for so many. We must try and avoid the ostrich burying its head in the sand syndrome if a truly compassionate society is to be maintained. Religions by their very nature owe us practical care at least as much as spiritual care.

Politicians and capitalists alike must show far more care for us all regardless of our class, status and race. We urgently need more care.

On the other hand; charity does begin at home and we must always bear this in mind. There is also something called tough love. Providing force in terms of policing is responsible, firm, and moderate - and in the overall interests of society - then it seems more acceptable than when it is unwarranted and unjustified. Punishment, too, must always be fair and just. Who would not respond to personal attacks on our loved ones with our inborn natural human aggression? Who is not incensed by mindless cruelty? However, we must never take the law into our own hands.

As ever I must strive for the broad picture - even if I have to play devil's advocate from time to time - as will become evident in future chapters. In order to show we care for one another we must, very occasionally, apply rules of conduct by reasonable force if necessary. We must rely on trusted authorities to care for us. If we let society degenerate to a situation where people are driven to desperate measures in order to ensure their own survival then we lose the ability to care for each other and further violence will inevitably ensue.

What is needed is a firmly caring society to minimise the damage done by the inevitable rogue elements in our midst. My message about the broader aspects of caring must surely be obvious. If this caring is within the formal social framework of human organisations which are traditionally on the side of good - then all to the good. These good organisations must surely lie at the root of our civilization and our consequent survival over a very long period indeed.

Some are more fortunate than others in the survival lottery of course. There is no accounting for this but there is much we can do by striving for better lives and caring for each other where we can. There but for the grace of God go you and I.

Inevitably human concepts of a caring God will also be further discussed (hopefully in an entirely open-minded and non-partisan manner) as part of the survival/caring debate. Tolerance, respect and caring are critical to the survival of a civilized society and for the making of better lives all round. If you live at all you are living part of one of the most wonderful acts of creation.

As compassionate, caring and conscientious human beings we have an undeniable and irrefutable right to our own beliefs. Freedom to believe, or not believe; as the case may be, is a highly prized social benefit. If you are atheist then you too have a strong (albeit negative) relationship with god anyway. If you are agnostic then you probably have no problems either. You admit you don't really know and are hopefully honest and open minded about it all. You can still care for humanity without putting any label on yourself.

Our urge to continue is convincing, compelling and compulsive. These are all beliefs essential to the very wonder that we exist at all - whatever the true explanation - which perhaps we may never know. This heartfelt belief in an inside force (the living soul) is a basic human right which no-one can take away from anyone. It has served society quite well for many centuries and may long continue to do so.

You are alive in your soul and no-one has the right to deny you this fundamental explanation for your own spiritual and physical welfare and your own right to survive accordingly. In due course I will suggest that sincere belief in a God of their own choosing: long before scientific explanations for natural phenomena, was of fundamental value to many people in terms of their survival. I will suggest that the concept of good: by whatever interpretation you may wish to put upon it, can only be preferable to evil for our survival.

In accordance with the age old rule of survival we swing between the opposing facets of our personalities known as Yin & Yang respectively. The rule according to Charles Darwin seems to be - adapt, adapt, adapt - in order to survive. In other words we must be open to changes, hopefully for the better, at all times.

Mistakes due to ill-considered leadership and misguided political principles, all too common, can be deadly for all concerned. Failure to get thing right in the past has led to untold misery for millions. We must learn from past mistakes to ensure future survival - despite some gross periodic blunders and conflicts due to mistrust and misunderstanding of one another.

I am going to try and show that some of the horrific and unforgiveable crimes against humanity have been ironically due to major faults in the very political systems and types of leadership we have progressively developed in order to ensure the mass survival of our species. This is indeed the greatest of all ironies. I will describe the worst of these aberrations as the _over-structure_ of society which is now on a worldwide scale. However, I can only speculate on man's intrinsically aggressive nature when family, territory and security are threatened.

33

I can clearly assume that, as a predatory species, we are capable of dominance, competition and appalling violence from time to time. On the other hand I can only wonder at the positive achievements of mankind in inventing religion, ethics and civilisation in general; where good normally wins out over evil to ensure continuity of our species. This is what I mean about caring. In fact this continuity has led to, not just basic survival but extreme luxury; for some of us at least. Many questions remain of course.

I may even question ideas of human leadership and how misguided leadership in the form of tyranny and bullying are a less than attractive part of our make-up. Perhaps in the present day the evolution of relatively liberal and democratic capitalism; for some at least, in the aftermath of some of the most terrible 20th Century wars of all time, has proven to be promising in terms of ultimate survival. This may well be a step in the right direction. This so-called liberal capitalism has even given some of us some reality, or perhaps some illusion, of freedom of choice.

Freedom of choice and competition seem to be the bywords of present day capitalism. How much freedom and choice we have in terms of commercial fashion, glamour and sophistication remains to be seen. Perhaps fashion is an outward expression of success and plenty? Drab utilitarianism is surely an expression of poverty, desperation and failure.

The classic quote in 1957 by Harold MacMillan (1894-1986) is as follows:-

'*You've never had it so good*'

This certainly rings true for many of us still to this day - and who would be a fashion victim?

Regardless of fashion we now live better and longer than ever and should be eternally grateful to be living in the present time as opposed to some less favoured previous generations. However, these good times may only be temporary and certainly apply to some more than others. They are always subject to change of course.

We can only speculate what the future will bring. For the time being our relatively high standard of living must certainly have, at least temporarily, enhanced our personal chances of survival for longer than ever known. We are extremely fortunate in so many ways. For now at least we are in the right place at the right time but should not become too complacent. It may be someone else's turn next.

The world has certainly come a long way from the good/bad old days when you simply took whatever you could get or starved. Or is this a passing golden era destined to eventual failure through the downright selfishness of some at the expense of others? Without seeming too pessimistic; one wonders if this is simply another political experiment doomed to failure - living on borrowed time - before running out of valuable resources.

We are all humans with our own characters and weaknesses and if nothing else the family unit teaches us to rub along with each other as best we can. We live in hope of a better future and this is all part of the survival conundrum. Pity we have such strong individual temperaments, opinions, differences and prejudices but this again must be a part of our instincts for a better world.

The overwhelming influence of our countless ancestors on our genetic make-up, in terms of both nature and nurture, must surely be at the very root of our mass survival to date. Thus, we simply are what we were made to be by our myriad forebears and their families too. To me this is what makes human survival such an appealing and addictive subject. Who has not wondered at their genealogy? Who has not sought to know their roots? It's that old curiosity again. Perhaps our survival is not such a neglected subject as I may have led you to believe.

Just think how many past family groups we are descended from. Surely everything they experienced or thought about did not just disappear into the ether? I shall talk of this later in a chapter dedicated to the idea of ancestral memory.

As individuals we are each just a small link in the eternal chain of human existence. This complex chain is interwoven with very many other human chains to form a multi-coloured tapestry of astounding energy, synergy and significance.

This is the very fabric of life, an inconceivable gift from our forebears: we owe it to them to make sure it remains in good hands. I have little doubt that survival can only really be assured where good surmounts evil. Is this why both sides of this dichotomy exist in the first place? Nature has a habit of pitting one force against another. If there was no evil conversely there would be no such thing as good. We must know one to know the other.

This is where human ethics, judgement and nurture are so vital. How I hate to play the preacher - it is bad enough playing devil's advocate. We must know right from wrong in the common interest. Again we must care for everything around us in order to survive. We even care for the animals we now eat. Yes, we are meat eating ex-hunters. We have learned to control our environment. We certainly are a race of exasperating contradictions and of conflicting intentions.

No one said why or when your heart must beat in order to keep you alive: that we each possess a soul; it somehow came about. This miracle of life in general is too sacred and valuable to be anything other than a wondrous gift to be nurtured preserved and maintained at all costs.

Caring for one another is vital for life to be sustained. We may not understand why we live in the first place but at least we can acknowledge every day of our existence as the most inexplicable and sacrosanct of gifts. Every day in which we survive is a bonus to be appreciated and treasured for all it is worth. Most of us simply take our personal, group and mass survival for granted.

Many think only of their own survival (and of the next generation or two) rather than the future survival of the species in general. We do not give such a fundamental idea a second thought in our day to day lives. This is somewhat surprising in view of the essential importance of the past, present and future continuance of our species. This seemingly sweeping and perhaps all too obvious statement therefore requires further analysis to verify just how true it may be.

Well we breathe, eat and sleep in order to survive for a start. We instinctively seek comfort, pleasure, happiness and affection all the time. We are a very social species. We are compulsive communicators. We control, compete, co-operate and interact with each other on a daily basis.

We are members of a number of differing informal and formal communities from our family to our nation. We are identified by numerous group headings: readers, writers, consumers, employees, pupils, pedestrians, voters, motorists, commuters, travellers, residents and so on. We love to combine into communal audiences, crowds and gatherings of one sort or another.

We even let our instinctive ancestral Jungian 'collective unconsciousness' take us over from time to time. We need to be together as individuals in order to survive. We are many things to many people. We identify with one another. Trouble is: we often form into cliques, or factions, in fierce opposition to one another e.g. pedestrians versus motorists (but that is another story). What are we like? Yet as a general rule, given the chance, we tend to care for each other. Thus we survive by loving, caring, sharing and adapting.

Let's take it from there. We are what we are. We are what our ancestors made us. It is all too briefly explained in a short foray into the extremely complex world of Genetics. Let us try and see what we can make of our ancestors. Does ancestral memory exist or is it just another figment of our all too vivid imaginations?

Try this for an awesome thought - In the split second moment of conception, the two streams of genetic information from your parents, handed on from generation to generation over literally hundreds of millennia, combined in one single cell that was to become you. It ensured that you became a totally unique living record of the lives and ways of your ancestors. And we are not just talking about the way you look - we are talking about your ancestral memories, the complete set of instincts and response patterns that were responsible for the survival of those two genetic streams in the first place. The instincts and response patterns that you were actually born with.

Dr. Bryan Knight

Theories of ancestral memory may well fall in line with the pioneering genetic work and laws of heredity of Gregor Mendel 1822-1884 as follows:-

Suppose one of your parents had the genotype AABB then you would have inherited AB from this parent. Suppose also that your other parent had the genotype aabb then you would have inherited ab from this parent. The combinations of AB and ab are parental types. Your genotype is AaBb and some of your children will inherit these parental types either AB or ab from you. However, it is also possible for some of your children to inherit new combinations called "re-combinants" from you. These are Ab and aB.

If you agree that we each carry the genes and stem cells of our forebears and that we in turn pass these on to our own children in parallel with those of our brothers, sisters and cousins; you may see this as the quintessence of survival. As each generation moves on to the next one I believe that the fundamental nature of our individual existence is logically replaced by the next one. My personal belief: that we simply live-on in our children, is borne out by the reverse idea that we inherit so much from our ancestors.

Just look at inherited traits such as the way we think, walk and talk. This is an ongoing process as each generation progresses to the next one. Each generation, of course, believing it is better than the last one (so it should be!). Conversely, the older generation often thinks it is better than the next – 'the youth of today'.

I am fascinated by mass group survival instincts inherited from our forebears. You need only observe flocks of birds, or pheromone (chemistry) driven ants or bees, to see how shared instincts, senses and telepathy make them all act in unison in such an astonishing manner - where the swarm takes over - a bit like the intuitive human Jamboree or Aboriginal Corroboree (mass gathering of the tribes).

The buzz of a modern music or religious festival may be a good example of a mass human fervour to be as-one? Do humans share a common hormonal bio-chemical life-force and/or electro-magnetic animal radar? Is this the mysterious *aura*? We can only speculate what this collective atmosphere, or electricity, really is in any mass gathering of living creatures. Do mass radiated live chemical molecules or electro-magnetic neurons combine in some way? A collective excitement takes over and individuals become as-one.

Many instinctively welcome this mass sensation of frenzy or togetherness. Why are some humans far more sociable than others? Some are dependent on others and love the crowds - others without being particularly misanthropic - hate and loathe the mass hysteria of crowds. There are loners and individualists of course.

Some join the mob willingly; others, mindful of their independence, do not like the idea of being taken over by the majority. The majority are probably in the majority. This surely can only be an inherited communal instinct shared by all intelligent species - including our own - all pulling together for the common good. Political systems, of varying success, have been based on this philosophy. At worst it is mob rule.

Until the discovery of the science of genetics it was entirely logical that one may live-on in heaven or hell as taught by religion. This continues to give comfort to many and is simply another survival instinct which is entirely reasonable to those fortunate enough to retain a powerful, enduring and fulfilling personal faith. Thus, whatever our religion, philosophy or personal values we all seem to share a belief in some kind of 'after-life'; for want of a better word. In other words we instinctively believe that we should survive in one form or another. A further example of this is of course the concept of reincarnation which has a remarkable affinity with the idea of genetics if you think about it.

Genetics were of course not known at the time of the formulation of most of the world's great religions but the power of ancestral memory, often known as instinct, has long been known. Most cultures have practised ancestor worship in one form or another as part of a wider set of values.

You have myriad ancestors from whom you have inherited many characteristics of race, personality and appearance. Like it or not their blood beats in your veins. Modern DNA genetic sampling techniques have successfully linked individuals over extended periods (even centuries).

If you are of European origin it is highly likely that at least one of your ancestors fought on one side or the other at the Battle of Waterloo (1815). Probably several fought at Crecy (1346) during the periods in question. The formula for the theoretical number of ancestors in a given generation is $2^n = x$; where n is the number of generations back and x equals the number of individuals in that generation.

Unless you have bourgeois, exclusive and undiluted so-called 'blue-blood' inherited from the aristocracy: I have to say that the rest of us 'commoners' are literally all genetic cousins sharing common blood and common ancestry. Sorry! Therefore, it is hardly surprising that all the hypothetical accumulated ancestral memory which may exist is so diluted, blurred and indistinguishable. Don't worry you do not have to remember all their birthdays. Yet, regardless of nobility, think what a powerful amount of residual and accumulated human life, love, experience and intelligence we have all inherited.

This huge collective latent living energy of all those countless human ancestors may at least serve to show us just what mass survival is all about. It seems inexorable, eternal and unstoppable and utterly, utterly compelling. So let us employ our inherited imaginations a little and try a few exercises in ancestor counting. The following antecedent calculations are hypothetical of course. These are merely to illustrate a point that we may never even have given a moment's thought to; such is the prevalent nature of survival.

Say you have two mean children - I mean 2 children on average. We may suppose that your real or imagined 2 mean children, as of now, have an average of 2 children each and so on (by squaring over an assumed 8 x 25 year generations) you will of course have 512 estimated progeny after 200 years. All these 'issue' will share their common ancestry with very many others of course.

Using this same very conservative formula in reverse in, say 2013, using the inverse square law and you too had a possible 512 great[8] grandparent's going back to 1813. Further, there was an amazing but purely hypothetical calculated 536,870,912* multifarious blood relations back in 1313. This in spite of the Hundred Years War (1337-1453), The Black Death (1348-1350) the Great Plague (1666), plus centuries of warfare, general privation, emigration and the appalling living conditions of so many lifetimes.

Now tell me that survival does not add up. This puts ideas of real human survival in a very telling historical context. Yet these are, of course, very approximate statistics to demonstrate just how very many theoretical compound ancestors/successors might have been possible, over not all that great a time; in terms of human existence. The reality is that past rural populations were so interrelated that these statistics are arbitrary to say the very least. I daren't go back any further. Remember we are talking literally of Survival equalling Life over Time as already discussed. Regardless of dodgy statistics; there has been a lot of survival over a lot of lives, generations and time.

This hypothetical figure over so many generations patently exceeds the entire population of Europe in 1313! This is an entirely theoretical figure of course. There were inevitable duplications due to small community inbreeding, numerous cousin crossovers and aggregate common ancestry in general. There were also non-reproductive lines to take into account. Probably numerous common ancestors abounded in close knit groups of villages. Many of our ancestors rarely travelled further than the nearest market towns. In fact many common regional characteristics such as high instances of blonde hair for example obviously date back to say Viking times. So the actual figures are not quite as realistic as they may at first seem. These hypothetical figures have merely been used to indicate the survival of so many of our common ancestors and the fact that we sooner or later must all be multiply and genetically inter-related one way or another.

Who says survival is not important? Most of us marry and if not having children of our own; certainly live-on in the genes of our blood relations at the very least. So what of our ancestors? Did we really inherit our inherent traits from them? If so, what is the nature of so called ancestral memory? The best word I can find to explain the influence of ancestral memory is instinct. We may have various traits in common but above all we rely on our instincts. These instincts may explain our often eccentric urges. Much that motivates us to survive, both consciously and un-consciously, is entirely intuitive and instinctive.

Theories of genetic inheritance, or ancestral memory, were known by Sigmund Freud (1856-1939) as 'archaic remnants' and by Carl Jung (1865-1961) as 'the collective unconscious' The term 'primal-instinct' also springs to mind when discussing our inborn intuitive impulses for survival.

I feel (instinctively) that we inherit more than just physical and racial features from our myriad ancestors. I believe we inherit mental and emotional traits as part of the genetic package. I believe that certain collective situations experienced by vast accumulated knowledge: particularly in awareness of danger and in terms of survival over many centuries, have been retained in our sub-conscious.

There must be at least some residual psychological blueprints for survival in our minds. Where does our instinctive fear of spiders or snakes, for example, originate; if not deep in the inherited psyche? Why do we fear the darkness, or the sea, or the weather so much? Yet we feel challenged by them at the same time. Such is the impulse for survival against all the odds. If we were neither daring nor cautious at the same time we would not have survived. Fear of the unknown is a powerful survival instinct.

Why do we both fear and respect our fellow man? How do we sense so many things around us? What is happening when we instinctively get up and walk to retain our equilibrium? Who is to say that we don't inherit other fundamental instincts? As already suggested, we need only observe our babies to see the whole wondrous evolution of mankind miraculously being enacted before our very eyes. No-one teaches that wonderful little baby to get up and walk.

There must be very many indeed of these intuitive survival instincts running through our veins. Many of these impulses or instincts must surely be based on a kind of collective mental state of mind inherited down the centuries. This is maybe what is meant by the word 'memory' (long term or residual memory) a long-standing subconscious perception, based on collective experience, which has become implanted in our genes to enhance our chances of survival.

This is a kind of latent survival manual which we disregard at our peril. Perhaps it all runs in the family as they say?

These instincts have surely come down to us with good reason in terms of survival in general. Many of us will no doubt recall strange internal voices telling us what to do or not to do at times of extreme danger or indecision. Remember the powers of imagination? There are many other overwhelming, psychological and physical survival impulses in our very being.

The best known of these is the instinctive electronic brain impulse, together with a corresponding surge of chemical energy, which makes us act urgently and without thinking, in order to save our lives. This would seem like a God-given boost to our chances of survival if we did not know something of the scientific principles involved. This surge of energy is like a powerful drug. Why does adrenaline make everything so urgent and vivid? Colours increase in intensity. We may even see red. The well known fight or flight survival mechanism kicks in. Where does that come from?

Does the caring instinct of our ancestors still manifest itself deep in our subconscious? The natural predisposition to care for one another may be just as relevant to nature as to nurture for all I know. There seems to be some credibility in this idea of handed-on human psychology in terms of survival techniques at least.

The term perhaps best suited to this strange feeling that it may all have happened before is *déjà-vu.* Those who have experienced various forms of mental illness, neuroses, delusions, hypnotism, superstition, dreams and drug usage for example may well have felt the real or imaginary power and depth of inherited or sub-conscious human thought.

There is a world of fantasy too. Certainly those of us who are inordinately sensitive to the powers of religion, superstition, spirit of place and fear of the unknown will identify with some of these ideas. A vivid example of what may lie in the folk-memory of us all may be sensed in the Celtic legends or the traditional sagas and fairy tales.

Music too takes our emotions to another place - a place deep in the soul - a place deep in the subconscious. Beethoven's glorious 5th Piano Concerto, which I am now listening to, is; hopefully, inspiring me as I write. I am showing my age a bit here. Whatever our age - where would we be without our music - be it broadly classical, jazz, or rock? Is this the meaning of the word: 'transcendental' - above and beyond? Has a piece of poetry, music or art never taken you there? Does one person's genius helps to encourage our own more humble levels of creativity?

We may be moved by the poems of Keats or Dylan Thomas, spellbound by the music of Mahler or the Rolling Stones and inspired by the paintings of Raphael or Matisse without quite knowing why. Is this proof that there is a higher level of human sensitivity and imagination: which we know exists but do not quite understand? Or is this just the creative right side of the brain, or perhaps the soul, at work?

This is something to do with communications of course. Music for example seems to be a kind of mystic language over and above mundane everyday speech. Dialogue is often inadequate for our spiritual needs. Listen to a great Requiem Mass sung in an ancient Cathedral if you don't believe me - then tell me you are still one hundred percent atheist (I believe most of them are ninety-nine percent in any case). Maybe it is more to do with sheer unmitigated 'man-made' beauty? I am sorry but I think, regardless of faith, we do all have 'souls'. I said we might not always agree. Anyway try it - or go to a rock gig - does you the world of good. Just listen to the voices.

I believe that human language, wonderful though it is, can be frustratingly inadequate when trying to express our innermost thoughts and emotions. I found this to be the case during the last paragraph. I very much regret missing out literature, language and communication (crucial to survival) in this project - maybe next time.

They say you can get by on a vocabulary of 1000 words but what a handicap that must be. For some, of course, writing can be more eloquent than talking and vice-versa. When reading we can become mesmerised by the flow of words but it is a silent personal kind of one-sided communication. You can't answer back: except by throwing the damned book on the fire. Please don't. At least when talking we can wave our arms about to give some emphasis to our meaning. We can even listen and that is the greatest of all things we can do.

There somehow seems to be this other spiritual dimension. There are many instances of insight, or paranormal, phenomena; which may be better explained as lying deep in the ancestral memory rather than as an actual living entity. The night is certainly a foreign country. Who has not been triggered into powerful visions of the past in very old, dark, sacred or otherwise haunting places? What are ghosts but an aberration of the mind? Why are they so deeply rooted in old and evocative places? Who has not felt moved by atmosphere? Scary!

Certain situations undoubtedly seem to trigger some kind of superstition, empathy or connection with our numberless past predecessors. Superstitious fear of the unknown is a powerful instinct. These intuitive feelings still seem to linger on even in our sceptical pseudo-scientific times. Hypnotism could perhaps release our natural inhibitions and take us back to an imagined or semi-imagined earlier time and/or place again somewhere lying very deep in the mind.

Who has not felt they had been here before? Here, as if by magic, we return to 'déjà-vu'. A kind of short circuit within our electromagnetic impulses seems to have been at work. Or was this just a trick of short term memory between the eyes and the brain? Like time-lapse photography. Or is this all just a state of mind?

There are recorded instances of profound visions, religious ecstasy and hallucinations which may well have their origins somewhere deep in the ancestral psyche. These are surely for serious study rather than mere supposition or superstition for the purposes of our discussion?

Spirituality in particular would debatably seem strongly related to ancestral memory and it certainly relies on historic myths and legends to reinforce this reliance on a long devout and distinguished past.

Do profound residual mass human emotions from the past somehow affect the electro-magnetic aura of very old churches for example? Or is the splendid old architecture simply evocative of a bygone era long gone and forgotten and never to be recaptured? It could all be an illusion and the product of an over-sensitive imagination or maybe there is just some remnant ancestral memory lingering on deep in our minds.

I have certainly been aware of a strange sense of *déjà-vu* on my travels in remote parts of Scotland, Ireland and Wales in particular. Many have experienced similar regression to past-lives when visiting centres of civilisation around the Mediterranean for example. What of Europe's glorious cathedrals and Asia's temples, mosques and synagogues? Your ancestors undoubtedly worshipped in at least one of them. This may help explain that awe we still feel and why we are so intuitively drawn to these breathtaking places.

Early Christian sites, for example, certainly have an overwhelming atmosphere about them: as do prehistoric stones, alignments and circles. Some believe in the presence of mysterious spiralling earth energy, resembling electrical eddy-currents, emanating upwards from the semi-conductor quartz particles in the stones - a bit like earth-leakage from a lightning conductor in reverse. The word *aura* again seems very appropriate to this idea. I wish I had more time to investigate these strange electro-magnetic phenomena.

My own view: subject to further research, is that the electrical emanations are more probably in our minds, as inherited from a long tradition of erecting megaliths over many millennia. Building temples to the Gods goes back as far as subliminal human memory. You somehow know instinctively that this or that holy-well or hermit's cell or ancient pre-Reformation Abbey was a place of pilgrimage to very many of our less cynical and more faithful forebears. Maybe the symbolic act of raising stones or burial cairns to the glory of whatever deity was being worshipped at the time remains a fundamental instinct for us all?

Who has not wanted to raise some kind of permanent monument for the appreciation of future passers-by? I recall planting a rowan tree - simply because I believed it to be a particularly beautiful tree. Shouldn't we all have planted a favourite tree by now for future generations to enjoy? Let these be our own humble monuments. We reach out to each other over very long periods of time indeed. We have survived for a very long time indeed and are very conscious of our own spiritual mortality. We make our mark.

Some say we all have a religious gene but one must be very cautious about getting out of one's depth. Many, who have lived comfortable indoor city lives, may find this idea of ancestral memory questionable to say the least. Yet we too have been drawn to the natural world from which we all originate. Why do we travel? We too may well have looked in wonder at the far horizons of our nomadic ancestors. We too have surely felt that irresistible urge to wander and see what lies over those horizons.

What of the intangible and awesome beauty of nature? We too have beheld the sea and the islands. What makes us so wonder at the sea? Why do we fear it and feel challenged by it: if this is not an instinctive feeling? How have we felt drawn to the mystery of the rock pools on the margins of the sea? Is this purely curiosity or is this a latent and ancient instinct to seek for food?

The writer recalls just such a primeval feeling with his family long ago on a very remote beach in South Africa by the thundering mighty Atlantic. Suddenly you were in this timeless zone. This is it surely - something primeval in us all. Or was it all in the mind?

Well we all came out from the sea incalculable aeons ago and its instinctive pull must have been fairly strong down all the millions of years from that infinitely remote epoch. The pull of the tides according to the gravitational pull of the moon is no less fascinating.

In passing the moon is the subject of a great deal of myth and legend. Folklore attributes lunar influence to both the menstrual cycle and lunatic behaviour for example.

The power of wind and waves is awe inspiring. No wonder the sea was a god (Neptune). So it seems likely that ancestral memories from relatively recent ancestors must still hold true. Do we seek shelter in the same instinctive way? Why do we feel so drawn to other humans - to be safely amongst our own kind? There is safety in numbers. Who taught us that? Why do we keep animals as pets? How far back does that go?

Instinct and ancestral memory go very, very deep. The sheer energy of it all can only make us wonder. Anyone who has been close to nature or at the margins of civilization will have little difficulty in believing in at least a residual ancestral memory. There are a lot of people living-on in our souls. Is it the sheer quantity of them all which makes all that residual living energy so confusing?

Remember the fundamental scientific principle that matter (energy) can neither be created nor destroyed? Einstein's famous 1905 equation - $E = mc^2$ demonstrates that (E) Energy, is equivalent with (m) mass x (c^2) speed of light^2, (where c = velocity of light in a vacuum at 299,792,458 metres per second). To exceed this speed is thought to exceed Time itself.

Thus ancestral energy, or Life-force, survives as long as the mass, or physical being, of the people survives through space, light and time down the generations. Thus, this fundamental equation remains central to my hypothesis that survival is related to the perpetuation of living energy and mass/matter, through space and time, as follows:-

$$E = mc^2 \qquad or \qquad SURVIVAL = LIFE\ over\ TIME$$

Thus we prevail. This seems an inescapable proof that we simply live-on from generation to generation. Both our common traits and our fundamental differences are ineradicable. They are chains binding us to the past and to the future. By chains I mean the mysterious and haunting spiralling chains of biological genetic energy. These have always been attuned towards mutual survival by our inborn selective breeding patterns: to be discussed in due course.

We must remember that we have very many ancestors with common genetic human threads still running through us all. We have seen the statistics. These genetic threads include that quality of being the most adaptable in order to survive.

Thus, there are no particular individual voices from our past ancestors: just a huge collective murmur. A bizarre, uncanny and unsettling murmur from very long ago - warning us, advising us, instructing us - helping us to survive perhaps. Or is this just too supernatural or over-sensitive to be acceptable?

Yet there is almost certainly an intrinsic, vague and disquieting truth lying somewhere deep in our subconscious. It is very difficult indeed to put one's finger on this ethereal idea. Yet who has not experienced a vague and uneasy dream-like feeling of *déjà-vu* from time to time? Odd.

Assuming this notion really exists: it can only be described as instinctive rather than rational. This is a mass instinct rather than an individual one. The inherited vigour and the blood beating in our veins, of so many people who lived, loved and survived before us is an awesome proposition, as stated in the above quote by Dr Bryan Knight.

Our very being sometimes cries out for the freedom of our ancestors - a kind of vague dream-like yearning to be out there - somewhere, now but long ago at the same time - somewhere timeless and eternal. I mean beyond the golf course, over the sands, across the sea, past the forest, beyond the hills. Not necessarily a fantasy in space or time, just out there?

The sheer subconscious power of the human mind never fails to astonish us. The untamed call of our remote ancestors still living deep in our minds must be very real indeed if we think about it. So is our inborn curiosity. The only appropriate word I can find in my thesaurus is *urge*: the urge to know where we come from and where we are going; the remorseless and pitiless human urge to survive; the strange urge to test ourselves to the limits; the insatiable urge to know.

Why do we travel? Why do we trudge across the mountains - mountains which get higher, colder and wetter with every year that passes? Is it better getting back than setting out? What is in the mind of the pioneer and explorer? No explanation, just a compulsive feeling from heaven knows where. Is it really all about the famous, or infamous, voices? These are our urges from long ago. Or am I mad?

Remember we are a strange mix of strange impulses (urges). Where do these thoughts come from? The writer has been fortunate, or unfortunate, enough to have travelled to some fairly remote places and can vouch for that old roaming instinct. In passing I am sure some of my readers will wonder just what on earth I am on about. Others may say: "Yes, I know those feelings". Some of us have more pioneer blood beating in our veins than others. Yet our ancestors no doubt fought in many a battle for survival. It's all in the blood. We are what we are.

The writer can also vouch for the need to settle down with a family but that old restlessness for freedom still occasionally calls. Difficult to describe - hands up if you know what I mean.

One need only hear a curlew or see the distant mountains and the sea to be drawn back to that ancient feeling. Heed the voices. Where shall we go on our next trip? Only when on holiday do we temporarily live the nomadic dream before returning to the relative security of our safe but boring jobs and our comfortable homes. But did we touch base just for a few short weeks? How sensitive are we to the voices?

Didn't the hot African burning sun on our shoulders; the sensuous calm golden mornings, deep blue sea, billowing cumuli-nimbus clouds, epic crashing thunder and lightning, the pelting rains of life and the magical following rainbows take us back, way back, far back: infinitely deep into primeval subconscious?

Again we are torn between the safe and secure side of our personalities (to which we have become adapted) and the wild and instinctive hardships of the lives of our distant old ancestors from long, long ago. Yet we know that, in spite of our inborn instincts; we would not have the skills to survive out there.

Well we have got used to our safe, comfortable 4 star en-suite lives and can now cover the planet at greater speeds than ever thought possible.

Despite having travelled a great deal in the aerospace industry the writer has never lost the wonder of flying. Why am I the only one still awake? I always, always look out of the widow. No matter how often you have flown, who with any imagination would miss the ethereal views of the awesome skies and the wonderful world?

In spite of our sedentary lives most of us have probably travelled more miles in one lifetime than our ancestors did in twenty lifetimes. See further comment on ancient voyages in search of new worlds. Now we comfortably cross vast oceans in hours; which took dangerous weeks, months and even years, back in the day.

In passing, never make the mistake of assuming that we in our modern world are cleverer or more resourceful than our cousins living in the wilderness (past or present).

We may live in a human jungle called a city but it is infinitely safer for most of us than the real one. Where does the warrior gene live in our psyche? Where does the instinct for the land come from? What of our love for the mountains? Adventures, dangers and experiences are certainly part of the human character for some people at least. Is this where the well known Celtic self-destruct urge originated? "What the hell!" warrior races often seemed more intent on the theoretical survival of mankind in general regardless of their own safety.

As experienced by the writer: the overwhelming call of the sea still remains since crossing it as a boy to see what lay over the horizon - the vastness of Africa too - something never to be forgotten. We may never be the same again once we have looked up at those unearthly night skies. Does the pantheistic search for beauty in the universe help to make us what we are?

Well physical travel across the topography of the world certainly widens the mind. We learn to lift our heads from that which is near, artificial and obvious: like this book or our television screens. We refocus our tired old eyes on the far horizons of our countless primitive predecessors. We are still nomads at heart. Why are the airport lounges, the freeways, the trains and the resorts so crowded? We roam in search of heaven knows what.

That big old world out there demands our respect for its stunning beauty. We escape the crowds of the cities. There are skies, deserts, oceans, plains, immense rivers, canyons and mountains. There are icecaps, volcanoes and tornados. There are things so big out there that we can only wonder at them. Above all - just for a moment: you've guessed it - FREEDOM!

What has freedom to do with survival? I don't know but if we are lucky enough to experience it then it must be good for us. What is good for us can only enhance our lives in general and our psychological well-being in particular. That very land and sea nurtured our very being as will become evident later in this presentation. Try it: go out. Get out of your vehicle. Get out there. Walk like a nomad but please don't get lost then your very survival could be in jeopardy! The spirit of place is there if we only open our minds to it. Life is not just about people is it?

What makes some of us gregarious people lovers and others introverted but outward looking? What of the visionary explorers and thinkers? We have many ancestors of many types and we can blame them for our wayward inherited characteristics. Are men more restless than women? Where did that innate, or inane, or insane, restlessness come from? Was it part of the great survival trek?

Does the restless search for farm and grazing land using ox-drawn wagons across the endless Veldt by the Voortrekkers, for example, not strike a chord somewhere in us all? Or is that old restlessness just a curse? After all what is better than coming home?

Home-making is, of course, the essence of both security and physical survival. As ever; these are just another example of the perverse human dichotomy inherited from our many, many, many ancestors. We have some very powerful urges to think about.

Some of us had some powerful juvenile impulses bordering on the obsessive when we were young: remember? What of our instinctive empathy as children for animals for example? Do these instincts originate from our far away herding background? Why are we so curious about things, so defensive of ourselves and families and so quick to respond to danger? Why do we still seek spirituality, explanations and answers? Why is so much of our make-up so natural? The words to truly express my meaning are not easy to find.

Let us use our overstretched imaginations just once again. Would it be too fanciful a notion to suppose that statistically you and I helped to create the wonderful Palaeolithic cave paintings at Lascaux in France? Do the maths. I mean as distant relatives? Don't these timeless images strike a common chord in us somewhere? If you are of European origins, like myself, then our common ancestors almost certainly were there in those caves so long ago.

If you are of African or Asian heritage then maybe your forebears too were there? What a fascinating thought. Nice meeting you, my cousin, again after 17,300 years and 69,200 generations - but please don't ask how many millions of bygone relatives we must have in common. Well, even excluding duplicated cousin exceptions it runs into many millions. We truly are all related. Surely ancestral memory (no matter how intangible and elusive) is more than just a figment of an over-vivid imagination?

There is something very deep going on somewhere. Our very sensitivity to the paranormal may well be explained once we understand our deepest of instincts and memories inherited from our many, many previous forebears. Perhaps we should just listen to the voices a bit more. Where were we all those centuries ago? Was it somewhere ancestral, deep, and evocative?

A large part of this essay on human survival has relied on personal belief in at least a vestige of genetic memory. It is more instinctive and intuitive than provable. Yet there is an irrefutable logic that the end result of so much natural selection, in order to survive over so many generations, has left us with a wonderful legacy of inner feelings.

Whilst I feel there is some credence to a residual form of elusive and indefinable ancestral memory (because I feel it in my bones) this is no proof that it actually exists of course. However, for now at least, I feel this to be an essentially controversial, intangible and indefinable subject best left to the experts or those with the well known Celtic 'second-sight'.

The vital influence of our present day family in terms of communal survival (if not always in the most harmonious manner) will be discussed in due course. For now let us look at some ideas on natural selection.

5 NATURAL SELECTION

I have called this principle by which each slight variation, if useful, is preserved, by the term of Natural Selection.

Charles Darwin

In discussing 'natural selection' the term 'selective breeding' will inevitably appear from time to time. I am not sure that I like either of these terms. There is something a little sinister about anything which even hints at genetic engineering yet these are entirely natural aspects of our survival. Temporary, semi-permanent or permanent pair-bonding, sexual contact and selective breeding are all critical to the survival of any species. We humans are no exception.

You may find this particular essay to be a little light-hearted and even tongue in cheek. The nature of the subject matter is often dry and academic but we can still relate some of the main issues to our everyday lives.

So far as is known we differ from some other animals by actually falling in love. This can be a permanent and enduring life-long commitment. This seems to be an emotional and intellectual concept which is not just peculiar to the human condition. It is evident that advanced mammals (and birds too) are capable of life-long partnership and affection. The concept of parental or filial love as opposed to romantic love is of course common to most of our fellow creatures.

Our human natural selection parameters go much further than just mating with the nearest biggest and best physical specimen. You know the one - the one with the glossiest coat, loudest roar and biggest muscles and so on. The ones with the highest opinion of themselves - the ones with the most perfect skin (very important - there is a multi-million dollar industry riding on this one) - the ones brimming with self-confidence and ready to take on the world. Just take a moment to look at your partner dozing in front of the television if you don't believe me. So how do we select our partner(s)?

We seek the ideal partner. Nothing less than perfect will do - the envy of all our friends - famous, urbane, clean-shaven, talented, rich, witty, slim, confident, warlike, virile, deep-voiced, rippling muscles, tall, tanned, fearless, good-looking, well-groomed, crisply ironed, immaculately dressed, perfect wavy hair, wonderful profile, smiling, humorous, sophisticated, in control, of noble Viking heritage as a bonus and above all - with the biggest whitest teeth in the whole night club - and that's only the women! 'Look at me the alpha male/female - just think what super babies we can make together'. Is this what we mean by selective breeding or natural selection, or is it a bit more subtle than that?

Human attractiveness comes in many forms. Sooner or later most of us will succumb to someone's unique qualities of body and mind. We may even be fortunate enough to meet our soul-mate. Remember the soul? Something we recognise as uniquely different and attractive in each other. We can't all be superior beings can we? These variations might be quite useful too. Our own distinctive characteristics could be passed on long into the future. Natural selection applies to the genes of course. The right combinations of genes will be preserved for all eternity for all we know. There could be a whole new world waiting for us.

This is not what we are thinking of course we are simply reacting unconsciously to the oldest survival trick in the book of life. We see it in their eyes, their smile and their every move. We are gone. We are enraptured. Who knows we could marry and have literally thousands of mutual descendants? What a thought.

This is what I understand by natural selection but I am sure there is far more to it than that of course. Intelligence matters of course. Integrity, morality and trustworthiness count for a great deal too. So does personality. Virility and fruitfulness, for want of a better word, must also count for something somewhere along the line. The affections of a lifetime should ensue if we get it right and we do not seek to compete for dominance over one another as so often happens.

Physical looks and a good physique are obviously important but many of us will naturally and unwittingly seek out Darwin's useful slight variation at the same time. A certain amount of imperfection certainly adds to the character and attractiveness of each other. Perfect human clones would be very boring, freely interchangeable with each other and would ultimately be very disappointing. They may even love themselves more than they love anyone else and that would never do would it?

In terms of so called selective breeding: a population of giant, lumbering and narcissistic Viking mounted warriors would do little to ensure our long term survival. Sooner or later they would outgrow their usefulness. Constant raiding, looting and pillaging would soon be considered to be a trifle selfish, extremely antisocial and very tiresome. Survival itself would become seriously threatened by the testosterone fuelled activities of these uncouth, hirsute and uninvited gentlemen.

Hooliganism is not something to be encouraged no matter how important selective breeding might be to creating a master race. See also comments on the word fittest to survive. The fittest are of course those most suitable rather than those who like Magnus the great hairy Norseman are the physically fittest.

Natural selection is all about the survival of the fittest i.e. most suitable. Most of us are extremely fit to continue the planet by the very fact that we are descended from generations of successful fellow survivors of all shapes and sizes. In other words: it is suggested that it is our very differences or variations which throw up new and experimental types of human beings; in order to make some progress in the world. Brains, character, compassion and personality seem to count for so much more than just looks and physical prowess - admirable though these may be.

We are attracted to so many things: eyes, hair, skin, a genuine smile, sense of humour, intelligence, posture, personality and so on. You need only read the lonely-hearts columns to find the right words. Above all this must be a caring person if we are going to survive long into the future. The importance which I place on caring for human survival is in fact written into the traditional Church of England Christian marriage ceremony as follows:-

I, (name), take you (name), to be my (wife/husband), to have and to hold from this day forward, for better or for worse, for richer, for poorer, in sickness and in health, to love and to cherish; from this day forward until death do us part.

In spite of the truth and beauty of these words: I am not proposing that everyone should rush out and get married and live happily ever after, of course. I am suggesting the importance of caring as part of mutual contribution to the overall struggle for survival. If marriage, or loyal partnership, is not about caring then what is? The huge importance of natural selection becomes more and more evident the more we think about it.

Honesty will not go amiss when selecting our partners either. Humour too will carry you a long way in the natural selection stakes. In fact it will be a vital commodity when times get complicated. The person, with whom we might spend the rest of our lives, needs to interest us and we must be interesting to that person. There must be a twinkle in the eye, not just a wicked one; an ordinary one will do for a start. Maybe this is why this particular chapter on natural selection is so tongue in cheek.

When we meet our prospective partner we must make an instant appraisal. We must read the signs and listen to the voices. What are our ancestors telling us to do or doesn't that matter anymore? Perhaps it is too late now: all taken in with one glance. We can be very perceptive of each other when choosing a mate for life and it pays us to be on our best behaviour at all times but no-one is perfect (rule number one).

Do we subconsciously look for one who resembles our parents; ourselves even; or our role-models, or just a presupposed ideal? Do opposites attract? Or does it just happen? If you know the answers to that one please send your answers on twenty three sides of A4 including a stamped addressed envelope.

Perhaps natural selection is just that: natural, something elusive but instinctive in us all. There is no doubt that we must get to know each other as well as possible before making that final and almost irrevocable decision to pair for life. There is too much at stake to make too many mistakes. I must reiterate that as an intellectual species we look a great deal further than just physical appearance when selecting our mates.

There is something brutal about strength alone. Although, I think I was a bit hard on those marauding Vikings. I have no doubt that great big hairy Magnus Bare Knees, or whatever his name, was a thoroughly decent chap once you got to know him. If we are of European origins he is probably there in us all. Magnus Big Genes lives on.

It is hardly surprising that natural selection as a concept is a very challenging one indeed. Yet in spite of past, present and future decadence I have little doubt that it will continue for as long as we survive to tell the tale. Pity so much that is artificial has now intruded upon our lives. We certainly do live in a fantasy world from which it is difficult to escape.

Many of us of course become role players according to circumstances. We learn, adapt and change all the time. Remember adaptability is a major tool for our survival as a species. If strong personalities do not compromise with each other then battles of will become inevitable. Elements of control and many other human strengths and failings may well be encountered in due course.

They say opposites attract. Yet the world goes-on and hope springs eternal. We are an optimistic breed when all is said and done. Most of us will moderate our behaviour to be socially acceptable but if this is merely a denial of a fundamentally flawed character then the truth will out sooner or later.

Some people are dedicated to keeping up appearances and will never let the mask drop. See chapters about Control and Social-Status. Others have more realistic expectations. Although we usually meet partners of a similar background and way of thinking we must never think we are somehow intrinsically 'superior'. We were all born naked. Some may well be attracted by people who are charming, exciting and dangerous. Well a varied, exciting and dangerous life on the edge may suit some more than others and of course such a life will never be boring.

Most of us, of course, go that extra mile to attract the person with whom we shall continue the world with. We fall in love. A set of blue, grey or brown eyes (the window of the soul) can be devastating of course. It pays not only to listen to the voices but to look deeply at the eyes when choosing our mate for life. Well we are selecting when all is said and done. When it comes to the voices or the eyes - then the eyes have it of course.

We are in fact giving ourselves over lock, stock and barrel to the needs of another. Yet most of us will have a hidden corner of our own souls which will forever remain our own. This might be a subconscious desire to keep a certain amount of mystery and allure within the relationship. We do not own one another after all. Or it may be to retain something of that certain unique variation mentioned by Charles Darwin. Only experience will tell whether we got it right. Personality of course counts too - but you don't need me to tell you all this.

We are attracted to one another, not just by physical appearance and personality - but by how caring, sensitive, loyal and affectionate we may be to each other. We are attracted to the person who may become a parent. We are attracted to the person with whom we hope to share the joy of children if we are fortunate enough.

We are attracted by the person who will pass on their genes, together with our own, to our children. This is natural. We want the very best of human attributes for our children. After all we will live on in them. We will nurture and care for our children and we will die for them if necessary. No wonder natural selection is so important.

Freud was certainly right when he suggested that the sexual drive was a primary motivational force for human existence and therefore critical to our survival as a species. When we fall in love, marry and have children we are simply practicing selective breeding to ensure the best possible chances for our family. Mostly, we select our own partners by one means or the other.

We may even conveniently marry the girl/boy next door. In some traditional cultures this selection is still done by the parents rather than by so called free-will. Either way, selective breeding is being practiced. Or nowadays we may well marry far outside our own culture. This may certainly aid the development of a mixed and multi-racial society in the interests of all concerned. Forced marriages of any description are of course utterly inexcusable.

Human sexual activity far exceeds the actual birth of children of course and the reasons for this are unclear. Some have a rich and satisfying same-sex relationship which does not of course result in bloodline children but nevertheless results in a rewarding lifetime of mutual care, love and affection.

Some heterosexuals too may regretfully never conceive and others may have to give up their babies. I am now a grandfather with a much loved family of my own but I was adopted as a child and am very sympathetic to this type of scenario. Others may go it alone and are to be respected accordingly.

Many non-human species have a specific mating season in accordance with the seasons or food supplies and we are more flexible and adaptable of course. On the other hand we are largely monogamous rather than random in our mating rituals. Perhaps our complex, unique but awkward physiques and postures made our mating habits different from the relatively straightforward mating practices of other animals.

There can be little doubt that we are a virile species. Perhaps the sexual bonding process within our partnerships has been vital for our survival in some physiological way as yet not clearly understood - caring? Maybe this is all part of the natural selection, adaptability and cultural changes deemed essential to our survival by no less an authority than Darwin himself. Certainly the parameters of Maslow's Triangle of Needs such as love/belonging do not belong exclusively to heterosexual or any other sexual groups for that matter. See chapter on Needs.

Talking of needs do we need to be happy to survive? The eternal human pursuit of happiness may well lead to a certain amount of disappointment. We do not have a god given right to happiness but when it does come along from time to time (as it will) it is a bonus and a gift beyond price. Savour it whilst it lasts.

However if we can achieve overall fulfilment, contentment and satisfaction in our lives we will be so much the richer for it. If we are cheerful, optimistic, enthusiastic, caring and tolerant by nature we will also reap the benefits in terms of happiness. I would simply suggest that humane, moderate and caring attitudes in general may well be potentially helpful to the survival of the human condition.

Perhaps one of the great satisfiers in life is centred round the family? It would seem logical to follow this quite interesting but difficult chapter about natural selection with further discussion on the vital role of the family to our survival.

6   FAMILY

Our most basic instinct is not for survival but for family. Most of us would give our own life for the survival of a family member, yet we lead our daily lives too often as if we take our family for granted.

Paul Pearshall

The family is survival at its most fundamental level. For the purposes of this essay I shall have to assume that the average universal family unit known for thousands of years is the one we have in mind. I am sure the principles behind my theories are all too obvious. Many survival issues require humanity, compassion and due sensitivity and this should go without saying of course. Like all social units, with the best will in the world, many things can affect the family balance.

I am only too aware that the ideal western family of say 2.4 children may be something of a myth. Families can of course be very much affected by separation, divorce, illness and so on. Nurseries, care-homes, hospitals, orphanages, schools and other caring institutions all have a crucial role to play in the wider contexts of family welfare. In all civilised societies there is some provision for those who need extra care. This role is gladly taken on in societies where the extended family is the norm. Indeed our own western society makes provision for extra family care in terms of grandparents, godparents and so on. The underlying importance of family life makes these problems even more sensitive.

There are two main types of family. There is the nuclear family and the extended family. Their names are self explanatory of course. Single parent families are of course just as valid as any other type of nuclear family. Divided, step or conjoined families may share common parents or even different surrogate parents and are every bit as important in terms of survival - sometimes even more so.

Adopted and fostered families again are just as significant. In fact all of these families are especially welcomed as part of the survival scenario.

Loving and caring are major features of human survival. These are found in every type of family or social group and families are both close but not too close if they want members to retain and develop their own individual personalities. In fact I must admit that these conjectural ideas about survival are of minimal practical value in day to day family life.

The above quotation suggests that we really do live on in our families and that our future survival as a species ultimately depends on this. The quotation also mentions how much we take this for granted. In other words the existence of our family, of whichever type, is the most natural thing in the world. So natural that it goes without saying and hardly merits a second thought.

The family must be subconsciously protected at all times and is sometimes the most conformist and most conservative of all organisations. I believe this innate family conservatism to be crucial to humanity in general.

The wider family unit of later years can also be the source of warmth, support and understanding. Mum and Dad should be there, where possible, if and when needed. That much loved cousin or aunt we all know may well be the source of a lifetime's inspiration and advice.

Survival is merely an underlying, unquestioned and unspoken aspect of family life. It is almost too obvious for words and therefore quite properly not given much thought to. There is usually a tacit agreement amongst families not to discuss issues which might lead to undue tensions. This is why we tend to take our families for granted. It is all so normal. It is only when a much loved family member leaves home for example that we begin to realize just how important the family unit really is.

We all hope that when our family members grow up that they too will have warm, safe and loving family homes. Yet those who are bereft of their own family will know what a struggle life can be and they will long to be fostered, adopted or become part of a caring surrogate family. Some societies even form close knit extended family type groups known as communes, offering an alternative to nuclear family life.

The family is more than just about love and shared blood relationships. It is about security, shelter, a safe home, food and all the many practical needs for everyday survival. Remember comments elsewhere about the many years of human childhood, care and education which the human species needs by comparison with other less intellectually developed species. This primary caring process can be something like 18 years at least and full human maturity may well take very much longer.

There is a theory that the natural love of parents for their offspring far exceeds that of offspring for their parents as the child gradually grows up and away from the original family - this being the natural consequence of human development. The difficult and delicate years of adolescence require empathy, sensitivity, tolerance and understanding as family members make that hesitant transition into adulthood and eventual independence.

Human development naturally tends to look forwards rather than backwards. If you think about the enormous love we have for our babies then you will see what an investment this love is towards the inevitable difficulties later in life when that original outlay of natural affection can be strained by the inevitable onset of outside influences.

It is only when we feel that love is threatened in any way that we react so strongly and problems of communication and differing values become so problematical. Even though we learn to let-go that initial love for our babies never really deserts us. Only they are not babies as such any more but this baby-love is so potent that it never quite goes if we are honest with ourselves.

This all-pervading maternal and paternal baby-love miraculously reappears when we become grandparents. It is said that the reason for the survival of humankind for three generations is due to the importance of grandparents in helping to look after the children throughout prehistory.

Technically, survival much beyond offspring bearing years is a luxury many species cannot afford. Survival of the species in general is obviously more important than longevity. It is only when they can significantly contribute to the survival of their families as leaders and teachers that the old will survive accordingly to the benefit of all concerned. In other words grandparents need to feel needed.

Our extended caring and educational needs require us humans to live for extended periods and in many societies the wisdom and guidance of the older generation is much respected. These caring family bonds may be subject to considerable outside pressures but remain of primary importance to the survival of us all.

The idea of caring, which has already been discussed, has its origins in the family unit of course. The family is the foundation stone of survival. Survival, nurture and learning; together with love, security and bonding, lie at the heart of family life. The family can only really function on these lines. The family may well be at the heart of human survival but is very much taken for granted as suggested in the above quote.

The hierarchy of the family will not usually welcome any unwarranted or theoretical ideas about how things should be done. The coherence and mutual co-operation of the family in general is of first priority and sacrosanct.

Controversy in families does little to keep things on an even keel. There is enough emotion flying around as it is. After all family balance is all important. It is more about getting on with things, even if unspoken, the underlying influence of the family on our very survival still remains critical.

Certainly many of the more inhibited, regimented and disciplined members of bygone generations were totally opposed to frank discussions about emotive or liberal subjects in any shape or form. One envisages the old ruling classes and the old conservative working classes in particular, being horrified by some of these new-fangled sociological theories.

Evolution, tool-using, survival, control, conflict, territory and so on - what do these contribute to getting the next family meal on the table? Yet if we stop to think about it: examples of all of the above may become evident around the family dinner table.

Society was and often remains too rigid, structured and conservative to countenance such liberal and controversial ideas. This is a symptom of the *over-structured* society due to over-population referred to from time to time during the course of this discussion. I am not suggesting that the family itself is responsible for over-structure but that it is obviously influenced by society in general.

Those who remember the post-war years may recognise the long overdue social changes that took place at that time. Did great-uncle Albert really wear suit, waistcoat, collar and tie on the beach?

Families are often about control when all is said and done. The family unit is both united and discordant at the same time. Generation differences, sibling rivalry, adolescence and divided loyalties can even cause bitter feuds which go on for years. Families can often be fraught with conflict of one sort or another. Why this is remains to be investigated further.

Perhaps families are simply too important to be threatened in any way either from without or within. Do families mirror the best and worst of society at any given time? Many modern families are grossly affected by emotional, financial, cultural and social difficulties. These outside influences do little to preserve the idealised family lives which we may wish for. Too close to home for many of us.

These ideas of family survival may be tacitly acknowledged but probably remain unspoken. Yet the family continues and the world goes on. We survive without even thinking about the reasons why this is. What could be a more striking confirmation of just how natural survival is to us.

You see families are more than just safe havens. Historically family units were often at the heart of tribal or clan genealogy. Some even have elaborate coats of arms and other symbols such as Scottish tartans to denote their regional family/clan pride and status. Some societies have traditionally been matrilineal and others patrilineal.

In many western societies the wife would and still does take on her husband's surname on marriage which would be passed on to their children. This was more than just a cover against the bygone stigma of illegitimacy - it firmly gave legal rights of inheritance and the like to the blood-lines of the family. Thus the family survives as a legal entity in terms of legacy inheritance and so on.

In Christian societies family surnames would often be based on inherited trades and professions such as Smith, Fisher, Miller, Carter or Farmer. Older trades or regional variations such as Fletcher or Cartwright remain to this day. Some surnames such as Tyler are derived from older spellings or misspellings of the trade or occupation. Many trades and businesses were of course passed down from father to son

Your superior status in society or your aristocratic regional origins may also be reflected in your surname e.g. Windsor or Mountbatten.

Even your village of origin such as say: Postelthwaite in Cumbria serves to show your Norse antecedents.

Your surname was often derived from your father's Christian name such as Wilkinson in England. Similarly Macdonald or Fitzpatrick or ap Thomas may show your Scottish, Irish or Welsh ancestry respectively. Magnusson may show your piratical Norse roots of course.

Some Christian names such as John or Mary were traditionally handed on down every generation either as first or middle names To go off at a further tangent and out of sheer interest in passing: I believe village legends of 'John Smith' supposedly surviving from, say 1563 to 1691, are most likely to originate from two, three or four generations of John Smiths rather than just the one. Thus was the urban myth invented? Although as a fervent admirer of born survivors I should perhaps be less cynical about these matters.

Christian names were often taken from saint's names of course and at this interesting point in our diversion to look at family names in some detail I shall return to the main narrative of this essay on families. It is important to remember that we are looking at the causes of survival rather than effects of survival. Yet family origins and their conservative characteristics are still crucial to our survival as a mainstay of civilised society.

Some of us have traditionally been deprived of the right to think for ourselves. This is an unfortunate consequence of our very necessary religious, philosophical, family and conservative political traditions of mutual control which have ensured we all know our place in society.

We now live in a more liberal world where ideas are freely interchanged by everyone on the internet if not in the family. This has opened many doors that were previously closed. Yet how strange that we still take such an elementary idea as survival so much for granted. Survival is simply an inbuilt and virtually subconscious instinct.

During the course of this project it will be seen that it is not so much the more obvious aspect of our day to day survival which I am discussing - it is more about our unconscious instinct to maintain the continuity of our species in the past and more particularly in the future. This continuity is centred of course around the family unit.

Continuity is the essence of survival. Continuity of daily life is therefore natural and unspoken to us. Life is both continuous and obvious. Life is for living. Life's too short to worry about things not immediately essential to our continuing presence on the planet. Eat drink and be merry for tomorrow we die.

As I write this essay on survival it becomes more and more evident just how obvious many aspects of our human survival actually are. Yet stating the obvious perhaps gives us the opportunity to widen our knowledge of ourselves - to open our eyes to the simple wonders of our being. Surely this is no bad thing? As the wonderful 18th Century humanitarian poet William Cowper says:-

God moves in mysterious ways his wonders to perform

One needs only to consider the lifetime love and care we give to our children and grandchildren to confirm just how true this idea of family continuity might be.

7   ADAPTING

In the struggle for survival, the fittest win out at the expense of their rivals because they succeed in adapting themselves best to their environment.

Charles Darwin

We have already encountered Darwin, of course, in terms of evolution in general. However, perhaps we should take a further look at this idea of adaptability?

This chapter on adapting for survival has been strategically placed near the centre of this exposition. What else ensures our survival? Who are the so called fittest or best adapted for survival? Well above all we are an adaptable, thinking and questioning species.

Thinking as individuals, or as groups, must surely increase our chances of survival. We would be nothing without our ability to think, imagine and remember things. Common-sense is a highly valued commodity. Having an open mind to a number of viable options is also of great worth.

Above all acquisition of experience by trial and error is fundamental to survival. There are teachers and pupils to ensure continuation of hard earned knowledge. We are an eclectic mixture of nomadic, warrior and sedentary beings according to how harsh the environments we find ourselves in and what genetic strains of our thousands of ancestors come to the fore. Who says there is no such thing as ancestral memory?

As nomads we must have knowledge of terrain, direction and climate in order to survive. As warriors we must be able to outwit our enemies to stand a chance of success. Like any hunter stalking its prey we must be extremely crafty, clever and quick thinking in order to outwit our dinner.

As farmers we must know exactly how to live off the land. We must take control of our environment. This is pure survival at work. This is our essential ancestral memory or instinct, if you prefer, at work

Knowing how to surmount dangerous situations enables us to live to fight another day. Finding means to defend ourselves in a hostile environment means that we must always be one step ahead of the world around us to even stand a chance of surviving.

Like all of our fellow living organisms we must have a constant and reliable source of food and drink for a start. We must be socially organised in family, clan and tribal grouping in the common interest. This is equally applicable to the jungle as to the city. Common interest can lead to survival or to conflict according to circumstances. We must communicate with one another for success. We must select suitable partners from a wide range of other communities to prevent inbreeding. No-one said it was going to be easy.

Unless we live in the tropics we must find warmth, light, safety and security in a hostile world of long, freezing and snowy winter's nights. Just briefly; go out on a cold, dark, wet and windy night in inadequate clothing; hungry and with no lights to guide your way and you will soon understand my proposition that our intelligence and natural instinct for survival remains of such importance.

Every one of us needs at least the basic essentials such as shelter, food and sleep for our survival. Perhaps I could add education, companionship, status and useful employment for our enhanced survival prospects. These are very fundamental human needs which we should think about. See chapter on Needs. We need to use our natural intelligence and intuition in order to survive and prosper. We must be adaptable at all times.

We need to know what we are doing, where we are going and why. We need to have aims and objectives. Some in our midst are blessed or cursed with an insatiable curiosity about almost everything they come across. Asking questions can only improve our chances of finding theoretical solutions to those things which might stand in the way of our continued existence. Thus we make some progress in our eternal quest for knowledge of ourselves and the World we all live in.

The Stone Age lasted for about 2 million years. The Copper, Bronze and Iron Ages spanned nearly 5000 years. The age of wind and water power lasted something like 1000 years. These age old methods were to be replaced by the advent of the coal-fired steam-powered mechanised Industrial Revolution of the 19th Century in the space of about 100 years. During the past 100 years or so during the 20th Century we have seen the mechanical, electrical and radio ages leading to our current fuel, transport, financial and communications era.

During this period and to the present day, we have extracted huge quantities of resources to manufacture our present day 21st Century super-materials such as: sophisticated metals, plastics, electronic components and anything else you may think of. These are the tools for survival at an altogether more sophisticated level than ever known.

How different from the lot of our poor ancestors who came on the long march from the Stone Age to the smog-ridden and dangerous Victorian gas-lit streets of not so long ago. They were of no real difference to us as people in essence. Less spoilt and perhaps somewhat harder working.

Nearer to nature, closer to the realities of survival they would have had little time to read this type of treatise on the ramifications of survival - they just got on with survival every day. How challenging their lives. How hard they worked. How little they got in return. How short life expectancy was for them. How prevalent was their child mortality. How limited were hygiene, medicines and food supplies for so many. Yet they too managed to survive by the best means available and often against all the odds. Remember their blood still beats in your veins.

How blighted their hopes and expectations were in the face of poverty and disease. Is it any wonder they turned to superstition and religion to help explain it all or to gain some hope and comfort when it was most needed? Unless you come from a long line of blue-blooded aristocrats or merchants you too will probably have had your origins in a very much less favoured world than you do now.

We all come from the land. Really most of us don't know we're born. We are very fortunate to be living in this period of history. Most of us have been spared from war, brutality, famine, starvation, plague and disease. Some of us have become immeasurably prosperous by comparison with past eras or even by comparison with those who still have to struggle to survive on a daily basis in the so called third world. Yet the third world is where many of the world's wealth producing natural resources such as oil, minerals and timber originate.

We are fortunate indeed to be so cosseted in life and (subject to conservation and more equitable sharing of our vital resources) we should make the best of it for as long as it lasts. The use of tools for the survival of mankind in competition with nature in general is discussed elsewhere in this presentation.

Until the advent of industrialization and technology, only 200 years or so ago, most of our ancestors still worked close to nature on the land. Day to day survival was still very much on the agenda. Believe it or not in spite of wars, famine and natural disaster it still is for many of us.

We have little concept of the realities of the survival and mobility of our ancestors for so many generations. In fact we have always grossly underestimated the exploits, capabilities and courage of our forebears. They were even more adaptable than we may ever have realized. Our seeming present day indifference to the reality of mass survival is due to a general lack of direct contact with our primitive origins.

We are often arrogant enough to suppose ourselves superior to less-developed people around the World. Yet we still form ourselves into various naturally protective and defensive social groupings of varying size and complexity.

The most basic groupings are either of family members, peer group bonding, military groupings, or educational classes for example. These groupings will by the nature of things change from time to time but are essentially disposed to serve the common interests of the group at any given time against any perceived threats from interlopers, strangers or outsiders. They may be fluid, adaptable and ever changing but they still remain as basic human survival units against a potentially hostile and threatening world.

Presumably these compulsory, or voluntary, groupings are primarily designed to create mutually beneficial bonding with one another. They primarily serve the common interests of their members. Mutual trust is paramount for success of course. This is all a natural part of our survival strategy. Strangers will only really be allowed entry into any interdependent family, group, team or gang when they become fully trusted by the majority.

It is not known what the optimum size of any family or other social grouping might be. The military squad may be of say 4 to 8 and these form platoons of 16 to 48. Sporting teams are limited by the rules of the game and space available for play on the pitch so are necessarily limited to say 4 to 15 players per side on the field at any one time.

However, there are of course potentially many more team members both on and off the pitch. School classes may vary from say 25 to 35 on average. These are all reasonably close-knit groups requiring loyalty and respect from the majority of their members in order to survive. They will all have both official and unofficial leaders and might be considered to be the basis of society in general.

These types of numbers, say around 25-30 on average may be neither too large, nor too small, to have a common consensus and to be of a manageable size. The writer recalls reading somewhere that the primitive nomadic family may typically be of around 25: this being the optimum number to ensure the welfare of all its members. There are many other temporary, extended and larger groupings of course but it can readily be seen that these basic natural human factions are designed for mutual support against all comers if necessary.

Every breath we take of course ensures our short term survival. We do not give this much thought. Why should we? At risk of being repetitive we just 'are'. What I am reiterating is how we subconsciously endeavour to ensure the long term survival of the human race in practically all we do.

As human beings we are, of course, very high in the hierarchy of the animals. We are even trained to think we are the best. Yet in an undefended aggressive encounter with crocodiles, sharks, lions or bears for example; we might well begin to question just who might be the best. Our long formative years have made us peculiarly weak and susceptible to harm in the wilderness. We are a naked and defenceless species really.

No wonder we are so often motivated by fear. If indeed we are the best then it was probably our very vulnerability which enabled us to surmount the rest by development of tools by which we not only survived but prospered as a species - tools as opposed to very big teeth. Give us the tools and we can do the job. We can build boats, wheeled vehicles, aeroplanes and houses. A few clothes too may help.

Now fully-clothed, housed, transportable and 'tooled-up' we can rule the world. We can adapt to anything. Challenge - bring it on. Thus we consider ourselves to be the most superior of beings: almost god-like, smug in the knowledge that we are top-dog in the scheme of things. With our infinitely adaptable tool-kit we could face any challenge.

Yes, we are an infinitely adaptable species who made it to the top of the tree of life. Darwin would have been very proud of us. It is all to do with success; success over all the odds, a very basic survival instinct. We call this the long ladder to success. However, if we really think about it, the means by which we reach the top are sometimes open to question.

Being adaptable does not mean that we are always the gentlest of people. We may be affable, adaptable and amicable but we may also be ambitious, aggressive and antagonistic at the same time. Some might even say we are just a piratical bunch of thugs, bullies and opportunists.

At worst we can be manipulative, controlling and devious. If this is what is meant about being adaptable in order to survive then Darwin did not exactly expect us to be angels in the process of surviving. Not merely just surviving but ruling the world: rich, mobile, selfish, safe, superior and in-control. Well some of us that is. There seems to be another inescapable conclusion emerging from the above discussion as follows:-

Like it or not - this is the nature of our species – truly adaptable but not always very likeable.

8   TOOLS

We shall neither fail nor falter; we shall not weaken or tire - give us the tools and we will finish the job.

Winston Churchill

It is obvious really that when the branch we were swinging through, using our unique simian opposable fingers and thumb, accidentally broke and became the first stick there was no turning back. We had unwittingly made our first scientific discovery. Here was our first tool, hammer, spear, club (golf or otherwise), lever, paddle, mace, badge of office, walking stick, drumstick, pen, berry-picker, cooking and eating tool, tent-pole, whip, shillelagh, truncheon, measuring stick, digging stick, shepherds crook, plough and so on.

Can you think of anything more versatile than a stick - ideal for nomadic or settled lifestyles? We could even envisage the concept of flying through the air using our simple stick as a throwing weapon. If we could do this with sticks what could we not achieve using stones - or both together? All we needed for permanent settlement. All we needed for shelter and safety in a hostile environment - even fire.

Our very adaptability and versatility ensured our success. So far as I know we are the only creatures to wear clothes. The invention of the humble bone needle enabled us to make skin clothes, boats and tents to conquer the world. Unless we become very hairy overnight how else could we have survived when our travels took us all over an often cold planet in pursuit of food, resources and land? Survival at sea demanded good clothing and onboard shelter.

Modern clothes are of course vital to our modesty and privacy. We can vary these according to climate, wealth, rank, fashion, gender or social status. One theory for our retaining only vestiges of hair in the present day and age is due to the development of an athletic physique and the necessity of sweating and keeping cool whilst hunting in tropical conditions.

Our skin colours, physical and racial characteristics would come to vary over many millennia according to geographical location and exposure to the burning sun. Our nomadic tendencies made sure we penetrated almost every inhabitable terrain around the world. We adapted in order to colonise the world. We needed clothes and shelter taken from the skins of other animals. We used nets, spears, clubs or harpoons to assist our hunting according to climate. I can safely leave you to think of all the many tools used by humankind over many centuries.

Stone and flint tools became virtually obsolete once metal was discovered and we could even help to build pyramids with them during the Bronze Age. We were nothing if not adaptable. We survived by outwitting our prey. We crept up on them, drove them, encircled them and trapped them. We had little choice in the matter. We are nothing if not flexible, versatile, artistic and inventive. We invented tools.

Would we have survived at all without these practical aids? I think we would have survived but back in the trees in the more limited manner of, say, the chimpanzees. However, the supply of trees and limited scope for hunting amongst them meant that sooner or later we would have to leave the shelter of the forests in order to find food and survive in a more open environment.

During the subsequent natural demise of African woodland and the vast spread of more open and drier savannah during previous episodes of climate change some ape species adapted accordingly to ensure their survival. Ours was one of these and other proto-human apelike species may well have developed in parallel with us but failed to develop as successfully as we did. Or did some species win the race for survival by killing off their smaller or weaker competitors?

Was it the chance development of superior tools which gave us the edge so to speak? Edge tools meant we could cut down trees, kill bigger animals and build boats. We could use edge tools as formidable weapons against competitors. This might explain our undoubted belligerence, mistrust and fear of strangers. It was all about adaptability and we were very good at this.

The replacement of dense forest by extensive grassland ensured the development of the grazing ungulates which were to become our prey. We had to compete with other formidable predators of course and again had to be adaptable and competitive in order to survive. We are talking of millions of years here: so it is hardly surprising that our survival instincts are so strong to this day.

Studies of fossil bones dating from the past 5 million years have shown how the familiar upright posture of Homo-Sapiens gradually developed. The head became situated more directly on top of the spine rather than at an angle appropriate to a stooped posture more suited to tree climbing activities. The feet became flatter for use on the ground and less prehensile than was required for tree climbing. The legs became muscular and athletic as required for living on the ground in open country.

The arms and hands previously necessary for tree climbing became adapted to the grasping of tools for hunting and for preparing food and shelter. The familiar upright human physique we now take for granted became more suited to running across the open plains (Bundu) in pursuit of prey. We were becoming more varied in our diet and were obviously becoming increasingly carnivorous in order to survive.

Due to a need for extreme fitness in running and hunting we required more protein than ever. Do you know that the Bushmen hunters of Southern Africa can still exhaust and outrun their prey?

We could, again by being adaptable, eat virtually anything we could find by foraging in the way of fruits and berries to supplement our increasing and insatiable demand for meat. We became nomadic and restless in the manner of some tribal societies still existing around the world. Maybe the other apes just missed their opportunities in our favour when the branch broke? Chance must have played a large part in our survival of course. We were lucky really.

Human technology was perhaps born on that fateful day when the branch broke. Not until we trained wild canine puppies to aid us in the hunt for meat or until we evolved the means to control territory to create farmlands and grow crops was mankind so destined for superior status in the animal kingdom.

The god-given supply of numerous types of plants to feed both ourselves and our animals did much to help us to put our roots down (literally). The unbelievable and miraculous invention of electricity and so on was a long way off yet but we had got off to a flying start. We have come a long way using tools within a structured society with common objectives. Pity it all became so competitive but that is the nature of things and we must live with this to this day.

Maintenance of territory and control rule so much of our daily lives. Maybe wealth and status comes a close second? How else could we have survived as a species?

In spite of our athletic abilities physically we are puny compared to some other species. Yet our brain power, spirituality and technology more than compensate for these shortcomings.

Thus we survived. It is said that when startled we instinctively throw our arms up and reach for the trees in the manner of our apelike forebears. We certainly react to unexpected stressful situations with the well know fight or flight impulse and the adrenaline rush that goes with it.

On the other hand; when we take the trouble to protect ourselves, by using tools to build suitable defences from predators and so on, we can see ourselves as a co-operative, defensive and cautious species rather than as an aggressive one. Equally attack can be seen as the best form of defence from other aggressive members of our own or other species. This dominance or controlling behaviour can be seen as an unlikeable but very necessary part of our human condition in terms of our survival.

We could use flint scrapers, knives, stone hammers and weapons of one sort or another. With some ingenuity; sinews, skins, wool or vegetable fibres, to use as rope, string and cloth, we could make clothes; build tents, ladders, boats, masts, sails, fishing nets, bows and arrows, musical instruments, defensive shields, roofs and so on. Clay for cooking pots, bone for needles, ochre for colour and even (eventually) metals. Now the world was your oyster to command at will. With your little red Swiss army knife with multiple blades and tools, all made from hardened steel, you could survive anything. You name it we could do it.

We were endlessly inventing ways to enhance our chances of survival and it is an ongoing process to this day. In spite of our own conservatism within a structured society we are all capable of the adaptability to change as suggested by Darwin. Through our intrinsic instinct to survive we were always learning.

Not only could we control animals for our own consumption but subject to climate and water supplies we could even begin to settle down a bit in our own territories. There is no place like home. Yet our need to maintain, fortify and defend these territories was to become a recurrent problem. What else could we do but join with each other in all our mutual interests?

Subject to the rule of law we could survive by living together. Families became clans - clans became federations - federations became nations - nations became states - states eventually became superpowers and the modern world was invented. We needed a wider less parochial outlook to avoid the dangers of inbreeding. Small bands of nomads needed to trade in their mutual interests.

Animals, tools and commodities needed to be bought and sold. Slaves, goods and materials would have to be exchanged. Perhaps before money was invented by the discovery of metals; other tokens of wealth and value such as highly polished symbolic stone axes, figures and idols were exchanged. Clay cooking pots, woven baskets and skin water bags would all have played their part.

Domesticated animals would have been used to carry or drag heavy goods until the wheel was invented. Thus markets were born. This process of mutual co-operation has resulted in vast cities.

The human city is like one huge mass of seething life like ant colonies or beehives. Should you live in one of the few remaining monarchies - we too have a 'queen' and (like worker-bees or soldier-ants) we all share in what is known as the 'commonwealth' to this day. The city is our hive. Not quite an empire but still a hive. Like bees we become one mass living organism over and above our individual and family needs - amazing.

This sharing is often very much less than fair to all concerned of course but at least we invented that most noble of ideas to ensure the survival of the fittest - civilization - and it was all because we used tools. Now we could climb the ladder of success and social status.

Independence? That's middle class blasphemy. We are all dependent on one another, every soul of us on earth.

George Bernard Shaw

I know that status is common to many creatures and we humans are no exception to this universal rule. In fact, if we are honest, it does seem to be a bit of an obsession with many of us.

We may not like 'knowing our place' but we seem to be stuck with this unfortunate instinct. So we try and 'get on' - upwards and onwards - up that ladder. We owe it to ourselves and our children. We send them to the best schools available so that they too can progress in their lives in due course. It has a lot to do with everyday survival.

It must be something to do with our place in an organised society all pulling together to a common goal. There are sheriffs and cowboys everywhere. Whether there are too many of one kind or the other remains to be seen. Society simply needs structure in order to live on.

Working with others goes far beyond what we may imagine. Despite keeping our distance from one another as individuals we are so communal as a species that we have become mutually interdependent on each other for our common survival to an astonishing degree. This required organisation according to relative value and thus was social status born. It is not a case of who is 'better' than anyone else of course.

It is now far more comfortable for many of us to live communally in towns and cities than living off the land or close to nature. A very few of us of us act like we are some kind of master race destined to inherit the earth whilst others do all the work. Most of us specialise in our own occupations and professions.

Despite a proliferation of DIY stores and garden centres we simply can't physically do everything ourselves. Possibly this is why we pay others to look after our basic survival needs. These others build our houses, grow and rear our food and for a price, cater to our every other need.

We buy our food ready cleaned and hygienically packed at the supermarket. Maybe we think it is all grown in a factory somewhere. Even the apples are of a standard size, colour and ripeness. Our high standard of day to day living is phenomenal. Do we really know the human price we pay for our high standard of living? Do we really care?

Or do we segregate ourselves from the realities of our inequalities from those who grow our food? We must never assume it is the supermarkets who feed us. They are merely the outlets. The farmworkers of the world are the ones who really feed us. We neither see nor think of them when enjoying our meat and vegetables. Who is killing our meat, baking our bread, catching our fish, growing our rice? Millions of tons of it - yet too many of us are oblivious to this fundamental fact. We are spoon-fed.

Yet those washing dishes in the kitchens owe their often meagre livelihoods to those at the top and vice versa. Their human credentials are no less valid than anyone else's. Yet unlike the primitive tribal network those at each end of the spectrum of our modern world will not even begin to know each other. The gap in relative status is just too vast.

What has happened is that we have developed mutually supportive and universal survival systems right across the vast and somewhat overpopulated world we all share. When these gigantic unbalanced rigid human control systems become too big, unwieldy, unjust and over-structured: then society loses control of itself. As you might expect; international conflict then ensues; as will be discussed in due course.

Thus our survival systems can become the very opposite when our leaders whom we have entrusted to ensure our mutual survival get it all wrong. When good is replaced by evil our very survival becomes compromised and I intend to try and show the importance of ethical and spiritual values to the civilization we all take for granted.

Through long evolved control systems, no matter how imperfect, everything is being done for some of us to enjoy our comfort and status. How did this come about? This is more than just surviving - this is all about life with all its inequalities and injustices as it has now become - yet perhaps here lie some answers to our dilemma. Is it this fundamental difference between us all as individuals which has led to our mutual survival?

No matter how morally wrong it may be: we simply can't all be the same or have our equal shares to what is available, can we? Yet that is no excuse for gross inequalities of course. I cannot stress this enough.

As human beings, we expect to be controlled, organised and dominated by others. This ubiquitous control is sometimes more or less subtle than we may believe and it pervades our very lives. We have succeeded as a species by being susceptible to control of one another. Or is it more about our never ending personal struggles for power, status and leadership to enhance our common prospects for surviving in a cruel and pitiless world?

We are simply a controlling species just like shepherds leading their flocks. Leadership, teamwork and adaptability are the answer to our continuing survival. These and many other questions will ensue as we further investigate the astonishing story of our survival against all the odds.

Over many centuries humankind developed various forms of government. With structured society we could enhance our means for survival. We could even develop mutually beneficent specializations, wealth, status and respect for each other. Class structure evolved so that we could progress, govern and control one another. When society became too large and sophisticated for family and tribal units it seems that control must have extended to forms of slavery, which of course we now find abhorrent but was perfectly acceptable at the time.

Despite, or even because we can be so well organised in general; we have had to face up to inevitable periodic conflicts between tribal groups, language factions, ethnic divisions and nations throughout history. The smaller the groupings are; then the smaller the conflicts between tribal factions or different language groupings. The larger the groupings are; the worse the conflicts. Warfare has seemed inescapable and we are an intrinsically contentious species when it comes to settling our differences.

We fight for survival, we fight for dominance, we fight for our rights and we fight for our very lives if necessary.

The all important achievement of status which must surely be so important to our species and others, of course, would not have been possible without a class structure. I apologise for any repetition about the significance of status but must stress just how crucial it is for survival. Yet relative 'status' is often an excuse to 'look down' on one another. This is the moral maze underlying ideas of relative status.

As with so much else: religion, ethics, morality, education and family values we need to feel good about ourselves. Again this bears some repetition: we need to feel good about ourselves in order to survive and prosper. It is to do with pride, respect and social value.

If you have nothing else your wealth, seniority, skill and personal standing (status) count for so much in the eyes of our fellow citizens. Those who feel unwanted by society will naturally form an alternative gang culture to gain status as an alternative to the official class structure from which they feel excluded. There is no getting away from status.

Whether we like to admit it or not we all aspire to some individual status. Thus we have all gone up a notch or two in the scheme of things since the bad old days of piracy, pillaging and slaughter. Well we plunder and pillage in a much politer and more managed manner these days. We are a race of managers. As Darwin suggests it is not how strong or intelligent we are: it is how adaptable we are that ensures the survival of the majority in the long run. This is not to say that strength and intelligence are not important.

On the other hand what could be less adaptable than rigid forms of authoritarian and militarised government: where our intrinsic desire for formal rank, position and status has overtaken our ability to think for ourselves? Certainly throughout the early 20th Century we had become over-nationalised, over-institutionalised, over-regimented and many ceased to have freedom as individuals.

The fact that the world now has so many middle-class communicators, administrators, white-collar workers and otherwise non-manual workers is proof of an upwardly mobile trend. This cannot go on indefinitely of course. Someone has to make things and grow our food. We now get others to do this. Such is our class system and hunger for social status. Service industries are now huge. Thus the world goes on.

Now we have wonderful medicines and the means for some of us at least to enjoy good, fulfilling and peaceful lives. We escape the worst aspects of the cities in our semi-rural retreats where we can live safe and contented.

As a matter of interest we may recall the Young Upwardly-mobile Professionals (YUPPIES) of the 1980's. These were the affluent new kids on the block who dominated the financial markets of the London City and some became flamboyantly wealthy as a result. Following a period of profligacy and extravagance many of these go-getting and thrusting young financiers were somewhat deflated following the stock market crash of 1987 and subsequent periodic recessions. Thus are the mighty fallen. There seems to be an inexplicable cycle of boom and bust.

The integrity of the banks and money markets were further seriously compromised by the global events of the early 2000's. Money had basically become corrupted by greed and dodgy investment practices. Capitalism that old mainstay of security, prosperity and survival was now becoming increasingly and alarmingly insecure due to ever more shaky foundations.

Money was no longer the exclusive preserve of the rich old aristocracy. Manufacturing wealth was being redistributed as traditional industries folded in some areas; only to be moved to others where wages were cheaper. We all saw it happening in the late 20th century and were seemingly powerless to do much about it.

Money was now becoming less exclusive. New money from often artificial financial investment and service industries for example was coming in. True wealth by making things was being compromised. What was being made was being made more cheaply elsewhere. New nations aspired to a share of the unequal world distribution of wealth in the past. Old money, in terms of inherited property and land that had once been fought for, had been far less of a gamble.

Money is a very elusive commodity. Money may now be distributed more widely and more haphazardly than ever but is still a force to be reckoned with. Yet the power of money now seems to dominate all social classes and the rapidly expanding middle-classes in particular.

Capitalism now pervades the mass media and the shopping malls as we all try to share in the good life previously reserved for the few. Greed and avarice too are now the democratic right of all classes not just the privileged few of the past. Many now stake their lives on winning millions on the Lottery. In my view money becomes devalued and corrupted both at the gaming-tables and stock-markets. I would prefer to see it more equally distributed but I am way too idealistic.

Some things never change: money = status = power. How fundamentally democratic this idea is remains to be seen. Yet the acquisition of money, status and power and the ambition that goes with them are simply part of the natural competitive human urge to survive and prosper.

We have yet to learn true democracy. Like ants we are now part of a huge collection of individual specialists, each doing their bit for society as a whole, each managing their own section, each at a different level in society: yet without always realising what the ultimate objectives of our communal efforts are all about. Unfair and divisive though this age old system may be: it seems to have been the only one which has ever really worked in past societies at large. Need this remain the case?

Can we restructure our society to allow a greater degree of opportunity for all regardless of our race, class or origins? I believe a start has now been made. With improved communications, education and liberal capitalism maybe there are signs of a better world for all emerging. Or is this just a passing phase?

Small tribal groups needed minimal social structures. Larger tribal groupings needed larger social structures. Whole nations, vast federations and so called Empires needed very many formal social divisions; such as the Indian caste system or the British class system. The Indian Hindi caste system consists of 5 basic hereditary levels such as rulers, warriors, merchants, workers and untouchables. Each caste has several sub-divisions according to occupation.

The British civil service consists of roughly 13 salary grades. The European aristocracy consists of an unbelievable 15 basic levels of hereditary nobility ranging from monarchs to baronets and earls. How we love our glittering honours. There are some big pyramids, big palaces, big egos, and big hats out there - and all that gold too.

In our universal quest for social status a proverbial situation of 'too many sheriffs and not enough cowboys' seems to have arisen. This is becoming a major problem around the world and one wonders who is actually doing all the work. Well, those with clean hands think themselves better than those with dirty hands. This is formalised in many societies. We now live in a world where clean hands seem to outweigh dirty hands. Even traditional manual duties can now be done from the warmth, comfort and cleanliness of a great big red tractor.

Some might argue that we are now largely governed from the City, capitalist banks and boardrooms where the real global power lies. These centres of financial power may be in Brussels, New York or Tokyo. However, each country, perhaps defined by race, language or culture, still elects, appoints or inherits its own type of political leader.

The fact that even our democratically elected leaders are often potential autocrats in the making still does not detract from this fundamental fact. The reality that some leaders are infinitely better than others is small consolation. A number of these despots in power still remain. Some puppet regimes have been encouraged by the west as convenient power buffers and as justification for subsequent invasion. When they get too big for their boots and bleed their countries dry is the time for their forcible removal - but not until then. Well better late than never. So how do these tyrants get where they are? This is social status taken to its extremes.

Man's historical inhumanity to man has done little to help a gradually emerging civilization based on latent kindness, culture, humility and humanity. A class structure of one sort or another seems to have been the inevitable consequence of our long march towards civilization.

For centuries these rigid class structures were set in stone and every level simply obeyed the orders of the levels above without question. Whether the orders were right or wrong you just did as you were told at all times. Often it was just an excuse for legalised bullying. Many did not even question these ruthless forms of control and expected little else in their own harsh struggles for survival in the age old manner and somebody infinitely better than you would always take the responsibility and make all the decisions for you. All you had to do was obey. Heaven help you if you didn't. Discipline with a very capital D.

Yet these mutually supportive systems undoubtedly worked for centuries when the hereditary aristocracy also cared for their so-called underlings in a paternal manner; often known in Britain as 'noblesse-oblige'. It was all about duty. Duties of mutual support in peace or war, in plenty or famine worked both ways.

However, by the time of the First World War following a century of Victorian/Dickensian inequality, industrialisation, insanitary slums, rural famines and gross overcrowding, things were changing in a number of ways. These were very much for the worst in some ways as well as for the best in other ways. Society had become ridiculously pompous, stuffy and absurdly *over-structured*. The rich had become richer and the poor had become that much relatively poorer.

We were to reap a grim harvest for these days of plenty and poverty. In spite of Victorian progress, devoutness and philanthropy: exploitation of the poor and dispossessed was still the norm.

The margins between compassionate charity and the dreaded Dickensian 'parish', poorhouse, fever hospital, lunatic asylum or workhouse became somewhat blurred. Read Dickens 'Hard Times' for more details of those times. Yet slavery, transportation, child exploitation and brutal forms of capital punishment were beginning to be seen for the legalised crimes against humanity which they were. Slowly but surely concepts of survival for all were beginning to be understood. The pioneering work of the medical officers of health in the 1890's following cholera epidemics for example were one of the greatest acts of humanity of all times. The growth of big cities required urgent action.

Survival again was at the root of these developments which we now simply take for granted. Fresh water supplies, sewage tunnels and railways were another Victorian gift which improved the lot of many. It was a time of mixed blessings for all. However the enormous wealth of some remained very much at the expense of others and was becoming increasingly untenable. Survival of the fittest was indeed becoming an issue. Those were bullying times and heaven help you if you tried to beat the system. Thus you went and died for your King and Country: in your millions if necessary. You simply did what you were told - no question.

By the first decade of the 20th Century the emphasis on royalty, nationalism and militarism were becoming remorseless and the Great War which ensued was an inevitable and tragic consequence. I use the word tragic lightly to describe the indescribable. It was one of the most horrendous, catastrophic and appalling crimes against the survival of so many.

The fact that it was followed only 21 years later by the Second World War in which countless further innocent civilians were needlessly sacrificed defies all understanding. As previously stated I can only surmise that society had become over-structured from top to bottom allowing the inception of brutal dictatorships. Humanity had lost all sense of reason.

When these formal divisions became militarised and mechanised, with 'God' and/or the King at the top and a vast plethora of ranks in between: then hugely destructive conflicts between vast Empire building nations such as Britain/France and Germany/Austria in the First and Second World Wars became inevitable.

Alternatively the Russian Revolution of the same period showed that poor people had had enough of the excesses of the long outmoded European aristocracy. This is not to say that what followed under Stalinism was any better. It was in fact infinitely worse; as is so often the case after a revolution. These were desperate measures for desperate times and did little to ensure civilised levels of survival.

Following the demise of Communism, the Vietnam War, the Chinese Cultural Revolution and the so called Cold War in general a more reasonable but uneasy world peace has now evolved. Now the vast majority of the populations of the west at least have enjoyed some boom years of post-war peace and prosperity with large scale diminishment of the old over-disciplined militarism in society.

Whilst remnants of these old imperial regimes still linger on; we now benefit from far more opportunities on the whole; regardless of class, race or gender than ever. The rigid traditional Indian caste system was technically banned at the time of Partition and Independence in 1947 but still functions to this day. Class structure has undoubtedly become more fluid and less multi-layered in general but still remains of course.

The expansion of so called middle class aspirations has been the result of increased spending power, consumerism and better opportunities for white collar employment. The old respectable and diligent working-class has become increasingly middle-class with the onset of modern technology, improved housing and home ownership. It is being replaced by a very different type of sub-class denied the dignity of even being working-class.

A kind of benevolent free enterprise culture has done little to alleviate our greed but has done much to improve our quality of life. Your everyday social status, or street-cred, if you like, is now sadly determined by the face value of your designer labels, mobile phone, eBook readers and the latest style in general; rather than by your exact rank in society. Watch they don't get stolen.

Everyone of course aspires to this newly found consumer led liberal capitalism and it is still a far from perfect system; where great inequalities of wealth still exist around the world. Despite terrorism, riot and insurrection our inevitable wars are now usually contained and controlled well away from the west.

Due perhaps to the underlying threat of nuclear weapons these inevitable conflicts have now taken on a different character as will be discussed in due course. These changes have resulted in better survival prospects than ever but are causing their own problems as the world gets more international; due to mass emigration to the cities and 'over-population' in general. As a general rule we have now become less racist, insular and nationalistic than we ever were in the past. Yet social status still counts for a great deal. What are we like?

Due to improved communications, transport and travel this mutual ant-like interdependence on one another has now become a worldwide phenomenon. Yet if the ant colony was destroyed a few survivors would regroup, form ranks and start the whole process over again. The instinct remains. If we had to go out and find water, catch our own food and build our own shelters then maybe the prospects for our personal survival would be far more relevant to our daily lives.

Instead most of us earn a living in one way or another by being (hopefully) usefully employed for the common good and ultimately for the collective survival of us all.

10   CONTROL

Which is better--to have laws and agree, or to hunt and kill?

William Golding - Lord of the Flies

The prevalence of social science in the late 20th Century covered a wide range of theories but with little study of survival as an underlying discipline. So why is there so little non-fictional literature on the subject of human survival? Has religious dogma and superstition somehow got in the way? These things are ascribed to the lap of the gods. Don't tempt providence. Previous more religious generations did not analyse things like survival as many of us are now prepared to do.

They were too busy trying to stay alive to think about survival. In any case it was more about the survival of your soul once your life had ended. Many still gain comfort from this idea. Perhaps we have never been encouraged to think for ourselves in these matters? I can only speculate that we are not seeing the wood for the trees. We are an often ambiguous, contrary and paradoxical species to say the least. Why can't we be more open-minded about life?

Maybe one of our most powerful human instincts is to be in control? We conceivably control ourselves and everything around us as a primary survival mechanism. We are an essentially controlling, warlike and predatory species; yet we can also be very caring and loving at the same time. We need to stay on top in order to progress onwards and upwards and thus continue human life indefinitely into the future. We have survived by taking control.

Can we be too controlling? Do class systems work? We are only really challenged as a superior breed by bacteria, micro-organisms and myriad insects. We are an intrinsically hierarchical species who perhaps invented a God-like image for mankind in order to rule both ourselves and many of the living organisms upon which we rely for our survival. Control is really all about being organised. Is it possible that we have been over-organised from time to time throughout history which has often led to warfare?

Imperial wars between ideologies, religions and nations in general have occurred throughout history. As a matter of interest why are the enemy invariably portrayed as inhuman and as belonging to an entirely different species? Do we really believe that our ways are so much better than anyone else's - that we hold the belief that civilisation - no matter how advanced - by God given right - somehow belongs only to those of a superior race or ruling elite?

Why did we tacitly let raving megalomaniacs like Hitler or Stalin take control of the lives of millions of innocent victims? Given the over-structured and universal patterns of discipline of the times what could have been done to stop them? The civilised world of centuries went literally mad. The Industrial Revolution had led to the deadliest mechanised weapons yet known. Never was human survival on such a mass scale so threatened.

Following the murderous, unnecessary and unforgiveable First World War and the (for some) generally self-indulgent 1920's something infinitely more menacing began to emerge in Europe and elsewhere during the 1930's. This was more than just an economical or political divergence. Capitalism itself was in danger of collapse. Human survival in the face of famine, mass migration and poverty was at desperation level.

Following the Russian revolutions of 1905 and 1917 and the great Depression of the 1930's a spontaneous backlash by the poor and dispossessed elsewhere was to be anticipated. Powerful right-wing regimes in Germany, Italy and Spain emerged and ruthlessly repressed the left-wing 'workers'. This created strange and ominous new pseudo-socialist parties, with racist and supremacist undertones, in an attempt to get the workers on-side. Control was the name of the game.

These undertones became repressive and pervasive and certain minority groups such as the Jews became scapegoats for the economic ills of society. Never had such evil regimes been known. Conflict between highly developed and heavily armed imperial and aggressive nations became inevitable. The fate of millions hung in the balance and nothing could be done to stop the inevitable carnage which ensued. The unstoppable major threat to human survival known as the Second World War then developed.

This was followed by various further post atomic-bomb ideological wars between rich and poor. Thus human survival has been more than compromised within living memory and we are left wondering how such an abnormality within a civilised society could have happened. Or could it ever happen again - god forbid?

Thus both Fascism and the much longer lasting but essentially similar repressive Communist mass control system originated before the Second World War. The latter survived in the USSR into living memory and still survives, at the time of writing, in North Korea. This was a fundamental dividing of the ways between so called left and right wing political factions and ideologies. This had already been seen during conflict between Royalists and Republicans during the French Revolution of the late 18th to early 19th Centuries to say nothing of a relatively minor and much earlier 17th Century English Civil War.

Society had become irrevocably divided by deep and sinister social divisions between rich and poor. This was nothing new. The origins of these social divisions may have been a deadly amalgamation of repressive feudal, imperial and industrial societies.

Control was exerted throughout due to a very organised but complex multi-layer class system where everyone knew their place. How else could an industrialised multi-layered society be controlled? All of these military, class and industrial systems were based on obeying orders without question down a very, very long chain of command.

Population had manifestly increased during the British Industrial Revolution and thus divisions between the various levels of society had become much greater than ever. As the differences increased between rich and poor; some kind of revolution became inevitable. In some places you nominally became either a capitalist or a communist. You had no choice. Yet both used the same oppressive methods. Control was absolute, brutal and most sadly of all accepted by the majority who knew nothing different.

Both Fascism and Communism ruled with a rod of iron. That is where the power lay. It was a society run by sadistic bullies. Not mere school bullies: they were demented imperial bullies. They were racial bullies. They were military bullies hiding behind their uniforms (just obeying orders). Many essentially good people were ruled by ruthless tyrants and their mindless lackeys. They were mad control-freaks without mercy.

Millions were routinely slaughtered and transported to death camps. You simply can't conceive or attempt to describe the appalling tragedy of it all. You become so upset that you can barely go on writing about it. What kind of humans are we? Yet in tackling a few ideas about survival you are forced to face up to some very grim realities.

Why was human survival so compromised on such an international scale? Why do we become so bitterly divided by gross inequalities and injustices? Does slavery and human misery as the ultimate result of control, bullying and exploitation in fact still exist? God forbid but it does: we know don't we?

I shall continue to argue that society had become grossly and dangerously over-controlled or over-structured. Have we not seen the dire results of centuries of an over-ordered society, where you obeyed orders without question, leading inexorable to the horrific international mechanised wars of the 20th Century? Only the hiatus of the Cold War and the hydrogen bomb, which threatened our whole survival, ironically resulted in a return to some kind of uneasy normality.

We are very privileged indeed to survive and prosper in our imperfect contemporary capitalist, communications, health, safety, mobile, consumer and hedonistic society rather than during the dreadfulness of previous generations. No wonder the concept of human survival is so vital. We must never take it for granted. We must always live to survive and survive to live. This is the fundamental truth of the matter.

Civilization has come too far to be jeopardised. Even if we take two steps forward to every step back we can only try to progress so that future generations can make a better job of it than we ever did. Hope springs eternal and we must hope for a better life; if only for the sake of our descendants - in whom I believe we all live-on one way or another.

Thankfully other potential survival characteristics of kindness, idealism and humanity are also available to us. These are often the catalyst for change. Thus civilisation can survive so that we all can survive. Survival of the fittest means the survival of those most adaptable to change. I have grouped ideas of competition and control together and it is easy to see how these two disparate ideas are related.

Competitive or potentially controlling behaviour can be either merely exciting or can be in deadly earnest. It is very hard to escape from competitive confrontations of one sort or the other as any chess player will confirm. In field competitions those teams or individuals who win are not just the craftiest or even the most skilful but typically those who are most physically fit. Conversely, in 'democratic' political leadership competitions it is those who are the 'fittest' i.e. most crafty, charismatic, eloquent and ambitious who usually win. In less democratic leadership competitions, the so-called fittest for command, can win simply because they are potential despots.

Power over others and the wealth that goes with it can be a very strong human motivator. What must it be like to reach the top of the tree in any organisation: to fulfil ones latent talents for organisation and control; to feel driven by ones-self and by others, to maintain hereditary class distinctions and/or competitive elitist school values? Alternatively a few have even made it to the top from the most humble of origins. The human lust for power can be ruthless and relentless. The bigger the organisation: the greater the scope to fulfil one's ambitions.

It could be argued that the modern world is little different from bygone days in terms of power and control. Again, is there still too much power at the top? Modern industry, government and commerce are believed to be in strong danger of becoming top-heavy. This makes for uncomfortable and unfair distribution of wealth but is very much in line with the middle class aspirations mentioned in the chapter on Social Status. The juxtapositions of power and influence are all-pervasive to this day. Yet ironically someone needs take charge or we would never have survived as an organised species. Perhaps it has all just got too big.

The dominant males of some species will ironically fight to the death in the competition for control, leadership and survival. This must surely be to establish the most virile leadership qualities to ensure the survival of the group in a harsh and unforgiving environment. The winner is often known as the alpha male and he has at least temporary top mating rights to the females. The downside of course is he must be ready to defend his rights from younger up and coming competitive males. This is not to exclude very strong females from dominance issues either.

Sometimes the balance between opposing factions results in bitter rivalry and enmity. A good example of this is demonstrated in the survival novel par excellence - *Lord of the Flies* by William Golding.

Who can deny that society seems to need a strong but fair hand at the wheel to ensure the survival of all on board? Someone needs to take responsibility. As ever, perhaps, a liberal compromise between inflexibility on the one hand and chaos on the other hand; seems the most desirable way forward? I may well suggest, in due course, that tidal swings between control and anarchy in the world have led to untold misery and bloodshed. The ideal would seem to be a neither over nor under-structured society based on rules in the mutual interest of all who are bound by them.

Who could object to reasonable rules of conduct applicable to all providing that these forms of control are essentially humane and reasonably democratic? An uneasy compromise between human dominance and co-operation has always been part of our survival strategy. The great American poet Robert Frost (1874-1963) usually had something shrewd to say about many things. This example is very typical, true and appropriate:-

The strongest and most effective force in guaranteeing the long-term maintenance of power is not violence in all the forms deployed by the dominant to control the dominated, but consent in all the forms in which the dominated acquiesce in their own domination.

However, this implied compliance between controllers and controlees alike has obviously done little to prevent unsavoury political bullying and domination. The latest tactics of world supremacy, human influence and control is of course to play-off opposing factions, nations or even despotic regimes against each other. Thus control is effected on an international scale. Some of these repressive dictatorships may previously have been actively encouraged by one side or the other when it suited their purposes.

Having seen the devastation caused by the mass systematic slaughter of innocent civilians in cities by 20th Century bombing new international control tactics have now been evolved. Either way, many of these formal conflicts are now conveniently fought in the desert in order to preserve our western democracy or so-called freedom. Thus our undoubtedly superior life-style is bought at the expense of others, 'over there', who are infinitely less fortunate than ourselves.

Or the hostilities are now more insidiously taking place on our own city streets in the form of terrorism, riot, revolution, anarchy, corruption, pollution, crime, gang-warfare, dangerous driving or other forms of everyday social unrest.

Some may even regard the potential conflict between haves and have-nots to be even more significant than the luxury of much vaunted green issues. Are these so-called green questions sometimes politically exaggerated to divert our attention from other matters; or are they a genuine concern for the survival of the 'planet'? How are we to stop the poor and dispossessed from chopping down the trees in order to survive? But are we really so capable of causing the climate to change so that this natural process does not prevail?

Are some of us jeopardising the welfare of all for the greed of the few? You might give up your car and holidays abroad and ride a bike in a well meaning attempt to save the planet but there are many, many more who want to give up their bikes (if they have one that is) in order to own cars and travel the planet. Maybe it is their turn now? We can hardly blame them. There are thousands of desperate immigrants trying to reach Europe to ensure better lives. Now what of survival?

Human survival for many is still relentless, tragic and often cruelly challenging and is all around us. Now tell me the subject of human survival should continue to be ignored. Yet mass migration is quietly and cynically acquiesced with; in order to cheaply grow, harvest and serve essential food supplies in particular. I am not talking of bygone slavery I am talking of modern times.

My research indicates there are now an estimated 12 million un-documented immigrants in the United States. Therefore, there is a universal and unlimited supply of cheap immigrant labour to ensure the interim survival of a very well-favoured way of life. They may not be slaves in kitchens any more but these poor immigrants are employed all over the world to support a lifestyle that they too would wish to be part of. Most of these first generation immigrants will of course aspire to better educational opportunities for their families in due course. If well paid work becomes available then this is no bad thing.

We do little to generate trust in each other in terms of social issues. It is often what our so-called News doesn't tell us that is really important. In fact News can at best only give a limited view of what is really happening. Reporting of major tragedies on television horrifies us but does little to help their future prevention. Ongoing less-dramatic items of equal importance are invariably overshadowed by major events. Radio is probably better for more balanced detail.

We settle down to be entertained or educated straight after the News forgetful that many are struggling daily for their very survival from ongoing and unreported natural and man-made adversity. Even our humanitarian feelings can be rationed and controlled by the mass media.

Some newspapers are, of course, still controlled by traditional press barons and politicians on the one hand and by popular customer and market requirements on the other. Maybe we should all be reading more books in any case?

Bookshops, libraries and literature in general seem to be comparatively free from outside control and long may it remain so. We can only welcome the increasing availability of eBooks to spread the word that books in general are no bad things in terms of democracy. We want no more historical episodes of book-burning: thank you very much.

On the other hand my family home is in imminent danger of collapsing inwards due to the sheer weight of books on the walls: so maybe a big bonfire is called for after all? No, on balance I would rather they went to a good home. Some are like treasured old friends. Yet here am I writing away to increase the world's endless supply of books.

What glorious independence: to be able to freely read or write with negligible controlling of thought processes. We just pick up the one we want to read. What wonderful things are books. I swear they have helped my sanity and therefore my very survival. So, no bonfires here.

Then what of the astonishing Internet? The whole world's knowledge is there at your fingertips and so quickly too: astounding.

At risk of sounding paranoid: what we buy, consume or even believe in all seem to be under some form of control. Many prefer it this way. We are being looked after when all is said and done - it could be worse I suppose.

Some have little desire either to control, or to be controlled, by others. Yet we live in a controlling society. When we travel and see our fellow humans living at such basic levels of survival we can only wonder how they do it. Some long overdue radical changes are now indicated as the common desire to survive on more or less equal terms gains ground in a world of distorted and controlling mass communications.

Power, conflict, manipulation and control seem to be an essential part of the human order. Are some of our power hungry politicians really the fittest to control our destinies and survival? Yet we would never have survived without some leadership qualities in defence of our perceived values. Humans are only human after all. We are all born equal. Some are of course more equal than others. Not only does this supposed superiority ensure the survival of our own predatory species but also that of other species upon which we rely in order to maintain our precious lifestyles.

Whether rich or poor and regardless of expectations, we all strive to survive; if not for ourselves then for our friends, families and children. It is not just personal survival either.

Whether we realise it or not we are intent upon surviving as a successful species. Some would call it simply control. Not survival of the species but control of the species. The natural human desire to live on, in one way or the other at the expense of others, is both powerful and inescapable. Our prime difficulty is our inborn selfishness in terms of survival. We all want to survive of course both as individuals and as a society. It almost goes without saying. All of us want to survive - it's only natural. So why do some have better chances of survival than others?

Here is the dilemma underlying this attempt at understanding the human condition in general. I simply can't understand why it is so much a matter of luck. You see you may simply have been born at the wrong time and place and maybe due to over-control you would have been caught up in all the horrors of human conflict or human indifference.

Why do we do this if we are such a reasonable and organised species? Are we over-controlled: is our society over-structured? Why do we take very natural principles of control to such extremes? Are we a good or a bad species - or a bit of both?

11   CONFLICT

What if man is not really a scoundrel, man in general, I mean, the whole race of mankind - then all the rest is prejudice, simply artificial terrors and there are no barriers and it's all as it should be.

Fyodor Dostoevsky, Crime and Punishment

I have read the delightfully idiosyncratic 18th Century classic *The Life & Opinions of Tristram Shandy* by Laurence Sterne and came across a classic quote about war by the eccentric wounded ex-soldier Uncle Toby as follows:-

"For what is war? What is it Yorick, when fought as ours has been, upon principles of honour - what is it but the getting together of quiet and harmless people, with their swords in their hands, to keep the ambitious and turbulent within bounds?"

So what is conflict? What in fact is war? This idea of war as a justifiable means of ensuring freedom certainly needs further consideration. How do "quiet and harmless people" get caught up in war? One is reminded of the range wars of the American West where far ranging cattle ranchers came into inevitable conflict with peaceable agricultural farmers. Is it to do with principle or territory or both?

Again I must put my devil's advocate cap on. For the sake of argument; has war been a necessary periodic evil of ultimate benefit to society? We need to defend ourselves from aggression at all costs - don't we? Not that it is always that simple of course.

As a species are we any different from other hunter killers? Our most aggressive tendencies and ability to defend ourselves seem to have got us this far on the long road to peace and understanding. Perhaps there is still a long way to go? No matter how abhorrent; has humanity ultimately benefitted from warfare? We are quite good at it - well there always seems to be at least some token warfare going on somewhere in the world. Conflict can certainly bring out both the best and worst of us and remains the foremost human enigma which it always has been.

I personally find warfare in general; incomprehensible, appalling and unjustified in a so called civilized society. However, I have mercifully been spared from it; so must consider myself very fortunate indeed. I have never had to test myself in the guise of a warrior. I am sure there is a warrior lurking somewhere in my genes or ancestral memory. See Magnus who has become the archetypical Viking ancestor, who we are both proud and ashamed of, elsewhere in this presentation.

Like any other predator, I would undoubtedly react aggressively to any perceived threat to my personal or group well-being - but a formal institutional war involving millions of protagonists for some impossible forgotten political ideology? How can such vast and inhumane slaughter be justified by so called religious god-fearing societies?

How can religious wars happen? Surely religions should teach pacifism and tolerance? Have we learned nothing? I certainly have no answer to these questions - especially in a treatise about human survival. I suspect this may be a long and difficult chapter as we grapple with the challenging and controversial subject of human conflict.

We shall approach the subject as open-mindedly as with every other aspect of survival; even though, from an intellectual point of view, every fibre of our civilised being cries out against warfare as morally wrong and inexcusable. When will we ever learn that our similarities outweigh our perceived differences? Yet it happens all the time in one way or another.

We have very different cultures dictated by our geographical and religious differences and conflict can easily be triggered by age old antipathies and misconceptions.

War is the ultimate and tragic expression of mass human hatred: a psychopathic loathing of unimaginable power and intensity. Hatred is the opposite of the human love and caring so essential for human survival as already discussed. Is it possible that you can't have one without the other?

When love is thwarted; then the equally powerful reverse emotion of hatred takes its place. Hatred must therefore be a natural human emotion which is usually suppressed in the interests of order but comes to the fore quite readily when provoked. We are often quick to take offence and defend ourselves and this too must be some kind of natural survival mechanism.

Most societies are multi-layered and the historical difference in relative status between a humble civilian slave and the emperor would have been equally vast. When relatively complex cities and trading centres evolved in farming areas the old simple tribal hierarchy was no longer viable.

As kingdoms, conquests and federations evolved the rulers naturally became ever more powerful. No wonder kings were seen and acted as gods in their own right. No wonder conflicts between god-like emperors were so universal and so bloody. If each had to obey orders from the level above then the pyramid of upward power was overpowering, overwhelming and overriding. Examples of mad and oppressive revolutionary emperors such as Napoleon or Stalin spring to mind. The ultimate sign of a megalomaniac is of course having power over so many other lives.

The fact that a blind eye, or even active encouragement, has been shown to these tyrants in the past does little to reassure us that we have backed the right side in the game of international politics. These autocrats are often to be found to this day as rulers in drought-ridden, feuding or backward states where a harsh, uncompromising and cruel environment has long dictated a more brutal and pitiless code of survival than we could ever imagine.

Many of these countries are plagued by constantly fighting warlike tribes, disparate factions and heavily armed revolutionary groups. They often live in remote and inhospitable territories of strategic importance to other adjoining territories. They become a natural haven for terrorists - although many terrorists like gangsters are to be found in our midst in the anonymity of our cities.

On the other hand some despots may control vast reserves of oil or vital minerals which enable them to play off competing western nations against each other at will - whilst supplementing their own Swiss bank accounts. Needless to say the west is often acquiescent with these powerful tyrants in their own interests and will similarly play off one tyrannical regime against another. I can now occasionally play devil's advocate when considering some ideas about those who actually go to war. As the vast majority of fighters are men I will refer to them as such.

Military service is an age old occupation which has been described as pure training and, in theory, offers career prospects, status, uniform and benefits. However, on the down side, it may well mean seriously going to war and should always be viewed in that light. Naturally discipline, patriotism, honour and comradeship are expected of all who take this major step. This is not something to be embarked upon lightly. Some are of course going to be better suited to the military life than others.

It must be recognized that some people are much keener on going off to war than others. However, military training with its characteristic mix of discipline and humour soon seems to lick us all into shape whether we like it or not. It has all been done before - sometimes with little choice in the matter. I wonder if we do really have a military gene in our characters. The uniform alone was a big incentive to a few past recruits.

The trouble is that natural primitive warfare was on a far smaller scale than it is now. Certainly generation after generation have followed each other in going off to war. However war spreads its insidious influence far wider than on the battlefield itself.

On the one hand; conflict can be tragic for all those who innocently get caught up in it. On the other hand; there may be some kind of nobility in this oldest and most hazardous of human occupations? No matter how civilized, pacific and reasonable we are as individuals we can still very readily admire some of the more estimable sides of human conflict.

Pride, courage, comradeship, leadership, loyalty and patriotism are admirable qualities by any standards. This presumably goes way back to our long, long history of pursuit, conquest and defence.

I suppose we are now looking a little closer at the *Conquistadores*. After the conquests people of spirit set out as pioneers of the so called New World. It seemed limitless then. Conquest from an overcrowded Europe was obviously needed as part of global expansion in terms of land, water and natural resources for rapidly expanding worldwide populations.

We became unwitting victims of our own success. We needed gold, copper and iron from Africa. We needed grain, rice and cereals from Asia. We needed land and more land from America and Australia. You did not win these commodities without a fight either. We needed fish from the sea. We needed more and more and more. We needed all the things many of us now take so much for granted. We now see these as essential for survival.

Our standard of living has gone up and our standards of survival have gone up correspondingly. We needed to conquer the world for its vast natural and human wealth before anyone else did. There were empires, commonwealths and dominions ranging from the immensity of Australia and Canada to the teeming plains and vast rivers of India and Pakistan.

There were British, French, German, Spanish, Portuguese, Dutch and Belgian colonies to name a few. Latterly the United States and Russia have contended for world domination.

We might call these a power balance which is giving many of us an unprecedented period of peace and prosperity. These are some pretty big and formidable players in the game. What of Japan, China and other Far Eastern countries - all striving for wealth and affluence? The balance is changing and who knows who will dominate the world in due course. Perhaps the nations with the biggest emerging populations will win out - who knows? Remember I am merely a commentator rather than a willing advocate for world domination by anyone.

We have now overrun the earth that supports us and these resources, such as oil and minerals are becoming increasingly rare as we use them up in order to maintain our modern technical, travel and consumerist lifestyles. Earlier I suggested that these natural resources seemed generous and therefore boundless but obviously this is not the case. Territorial issues are obviously a major cause of conflict. I wonder if a quite simple conclusion about the prevalence of human conflict is now beginning to emerge from this two way discussion.

Discussion, by the way, is always the best way to resolve our differences. Maybe human survival in a harsh and difficult world, since time immemorial, has imbued us with a very natural fighting spirit: breeding qualities of courage and hostility without which survival would have been impossible. In spite of civilisation, or maybe because of it, this latent aggressive energy is simply too powerful to contain? It is in us all. We are human beings.

From schooldays we are programmed to become winners. These days our more belligerent, braver and adventurous individuals will often succeed in competitive arenas of mock conflict such as politics, business, sport and management. Many young men in particular are naturally aggressive and this is something easily forgotten as we become older and more cautious.

When this aggressive warrior gene is frustrated by childhood abuse, unemployment or environmental issues for example then anger and communication problems in adolescence may cause significant social problems and self-destructive urges later in life. In a world often deprived of adequate employment, physical activity, manual dexterity and creativity these problems are common to many frustrated individuals regardless of their origins.

It is interesting to note that when rich are pitted against poor; the superior firepower of the rich is invariably compromised by guerrilla tactics, or terrorism, in which the war becomes prolonged and increasingly bloody. Language, financial, class and consequent cultural differences are traditionally causes for resentment and suspicion. I can also see how poorly thought-out artificial boundaries and federations of traditionally disparate and opposing nations will come into unavoidable conflict sooner or later. Many of these unbalanced federations have their roots in archaic and divisive ideologies inherited from long defunct regimes.

Political mistakes have a habit of lingering for many decades or even centuries. All wars are of course one huge and tragic mistake or a case of damage limitation at the very least. I can also understand how relatively powerful, rich and prosperous societies become greedy, manipulative and autocratic. I can even see how bitterly divided tribal or disparate religious communities may well be at variance for generations due to perceived ideas that one faction may be 'better' than the other - each having been mistakenly taught that theirs is the one and only true faith. There are some obvious and well known examples of these intractable problems.

Many other animals will be ruthless predators in order to survive yet will only seriously fight each other to death on rare occasions. Despite often horrific wounds they usually back off from each other once the pecking order has been established. On the other hand some of them may compete for food and territory in packs just as we do where the individual free-will does not come into the equation. Thus you get pecking order against pecking order almost as if they belong to different species. The mob instinct takes over and fights to the death can take place.

In modern technological warfare we tend to fight each other from a distance. Thus our killing is less personal. It is almost a fantasy kind of warfare like some kind of computer game. The realities are somehow mitigated by distance. We fight an unseen and anonymous enemy. We may be remote from our opponents yet the killing and dying is no less barbaric than it was when hand to hand combat was the horrific norm. Mind you we were far less sophisticated in those days.

Examples of almost unbelievable and fanatical bravery and self sacrifice are not unknown amongst some traditional warrior races to this day. It is as if the minority are willing to sacrifice themselves for the majority. Thus altruistic self-sacrifice may be seen as some kind of survival strategy. We don't personally want to be amongst the few of course but somehow realise that this might be required of us one day.

What are any one of us capable of once our blood is up? Our pride has been challenged. We are angry beyond control. The dormant killer in our ancestral memory has taken us over. The provocation to our natural pride has become unbearable. The adrenaline in our blood is very real and powerful. We have been around this hostile place for long enough to know what we must do to survive in it. We must have stamina, fortitude and determination to succeed. We must be strong, brave, clever and resourceful to gain the respect of our fellow human beings.

We must also be seen to be tough enough to deter anyone from messing with us. Thus we swagger, bluster and posture accordingly. There can be little doubt that many of us are descended from a long line of heroes and protagonists. We are programmed to fight to the death if necessary. It is all a huge gamble of course. It is one I can barely comprehend - but true.

Man is capable of horrific crimes against his fellow man and one wonders if indeed we are capable of becoming an almost separate species when we go out to fight each other without mercy or reason. It may be dormant but is still there whether we like it or not. Man the subjugator inevitably becomes man the protector. We are then fighting for our very survival and as already discussed at some length; survival is the most powerful of instincts. This idea of fighting a supposedly sub-human enemy illustrates the frenzied propaganda to which we are subjected in order to justify the mass hysteria and killing fury of modern warfare.

Our innate nationalism and fear of strangers: with different values, languages and perhaps religions, is soon triggered. We begin to suppose that we are on the all important side of the good which we were taught about as discussed previously. We are sincerely led to believe that we are intrinsically good and the enemy is deemed to be fundamentally evil - which may sometimes be true for all I know.

One need only look on the proud faces of the marching veterans to understand how as young men they were tried and not found wanting. I might even do things in war which I could never have believed myself capable of under the discipline, stress and terror of it all.

Is warfare really natural after all? My rationality would probably have been taken over by the fighting genes inherited from my more savage forebears. In the heat of battle the power of the internal drug called adrenaline (and that of my comrades) would switch into the mode of the pack lying deep in our ancestral memories. We would kill together in order to survive together. This would be our survival against theirs. We would not worry too much about mass human survival until later when the history books were written and the folly of it all became apparent. It would be us or them, there would be no choice in the matter, do or die.

We might even revert to praying for the mercy of our inherited 'God' under these desperate circumstances. What would we have to lose? Thank the good Lord of mercy (if there is one) that I have never been tested out in this matter. Ask yourself this: what would you have done? What would I have done? We are no different than them are we?
Perhaps joining the armed forces, when young and eager for travel, adventure and comradeship is considered well worth the risk for those with few alternatives - they even look after you in a sense.

Someone has to do it and many are more than willing to test out their own courage and endurance: to face up to danger and prove their worth to themselves and others. Who has not asked themselves how they would perform in the event of being caught up in bitter armed conflict? Fight or flight? Not much choice is there? We must obey the order to fight whether we like it or not. It's us or them. It might be better if we did not come into hand-to-hand eye-to-eye combat with a man with whom we might just enjoy a glass of beer and take pleasure in showing each other photos of our lovely wives and children.

My lengthy analysis of some aspects of human aggression and its relationship with the concept of survival has again merely served to show what an ambiguous and puzzling phenomenon it really is. Mercifully, I was spared from all this – many were not. I have been obliged to play the role of devil's advocate in order to try and understand something which I instinctively find truly abhorrent, inexplicable and barbaric.

Despite the length and depth of this chapter on human conflict: there seems to be more questions than answers once again. If any conclusion is to be reached I suppose it is quite simple really. We are a peaceful species in general but a fighting breed when roused.

Some are more easily roused than others.

All men by nature desire knowledge.   *Aristotle*

In writing of metaphysics, life-science and religion I shall unavoidably have to resort to some personal beliefs and disbeliefs. What else can I do? If we think about it; these subjects are mainly down to our own cultures and values.

For example how we think we might live-on after death is mainly down to our education and individual beliefs. Then we become divided broadly according to either religious or scientific principles. Some neither know nor care. Others fervently believe that the individual soul lives-on in heaven: that our lives have been not only worthwhile but more importantly, worthy to be returned to our maker. This is a natural, spiritual and comforting thought. Others more pragmatically believe we simply live-on in the genes of our family and successors.

As a free-thinking realist I instinctively incline to the latter view rather than the former. Either viewpoint shares the common values we all place on life as being somehow everlasting: too great a gift to be diminished in any way.

So whatever outlook we might hold: what if we look at things from a *metaphysical* point of view? I see metaphysics as a simple tool for exploration of the realities of life and survival. We may define metaphysics as being the study of the fundamental nature of being and the world: so it is very relevant to ideas of human survival. Metaphysics asks simple questions about life and candidly promotes free-thinking, clarity and truth.

We might benefit from looking at things from a somewhat different perspective. Let us open our minds to a new way of thinking. Let us try some very basic metaphysical questions about survival. Q. Do we live-on? A. Of course we do: in our successors. Q. What is the Life-force? A. The mysterious Life-force maybe lives in our very souls. Q. What is the nature of life? - Is it electro-magnetic? Is it chemical? Is it biological, or all of these things (with a dash of mysticism) and more? A. I don't know. What is life? A. You tell me.

Perhaps the ancient philosophy of metaphysics may give us some further clues as to the nature of our very being? Just ask yourself some very straightforward questions. You may not get many answers but you will have learned to turn things upside down to get at some of the underlying truths.

Shall we look a little further into the possibilities presented to us by our new metaphysical way of looking at things? Shall we try and analyse a few hitherto little realised concepts? I for one would certainly welcome being identified as metaphysical, or maybe as a realist or even as a pragmatist and so on. The connotations are endless: we are many things to many people.

We are ruled by our *isms*. How spiritual I may be remains to be seen. Thus I am not an evangelist, more of an evolutionist. I am free to think for myself. It may be an old cliché but I am free to be me. I can now ask myself all those eternal questions. For example my reading on the world's great religions and ethics in general leads me to suspect that humans might be genetically programmed to be spiritual, sensitive or receptive; regardless of any professed faith or lack of faith. Is this the clichéd religious gene?

I see the whole area of spirituality as being strongly linked with the powerful human imagination. Thus we can envisage those things which we cannot necessarily prove. Human qualities of humanity, loyalty and fidelity are important aspects of faith; yet these are often little thought about by their more vehement detractors. I personally suspect that more extreme aspects of atheism may well be just as bigoted and narrow-minded as the more extreme aspects of religion which they so abhor. Why is this?

Regardless of atheism and agnosticism I do not see why metaphysics should be thought of as being opposed to the philosophy of religion or spirituality in general. It simply asks some pertinent questions rather than relying on unthinking or unproven convictions. It is pure imagination. It is the essence of intellectual freedom. It is like a fresh wind blowing away all those cobwebs. New doors are opened. New horizons open to us. We feel liberated from all that we have been led to believe. There is nothing quite like it. Try it.

The dream pursuit of that glorious and elusive product of the imagination; freedom, never quite leaves many of us. Yet many of us will never ever be free. Many are totally hidebound from the cradle to the grave. Can this be right? We are all fellow humans after all. Yet some are just born in the wrong place at the wrong time. They have been taught to do as they are told. They are taught not to question anything. They must know their place. How sad. Some of course have more imagination than others. Above all many seek education in order to escape from their plight.

We all surely have aspirations, hopes and dreams regardless of our relative status. In fact those who would sell us these dreams are only too aware of this fact. Human experiments in depriving workers of freedom, education and imagination, as part of political totalitarian regimes during living memory, also come to mind. This may just have worked by the use of slave labour when building the utterly useless Berlin Wall or the priceless, if slightly over-engineered, ancient Pyramids but maybe not when constructing a highly technical and extremely useful, if controversial, nuclear power station. This requires much thought.

Do my own metaphysical/survival theories fly in the face of long established taboos and general narrow-mindedness consistent with long established human control systems? Are we really so controlled and controlling that so many of us seldom think for ourselves? What label must I put on myself to gain some credibility for my thoughts on these matters?

The truth of the matter is that I love words. Words are the ultimate expression of the human imagination just as the eyes are the windows of the soul. We just have to be receptive to them. They are communicators vital for human understanding. Thus, if I am lucky, my imaginings about my chosen subject may perhaps stimulate your own imaginings and existing knowledge of the given subject. Words are creative, universal, descriptive and evocative. At best they are like brush strokes or music notes. Used correctly they can be pure poetry; if we can only find the skills to put them in the right order at the right time.

Words are all about communication, ideas and knowledge in general. I believe there are about 61,000 of them alone in this short book.

Next to art and music; literature is my greatest intellectual pleasure. I also have a strong interest in psychology and philosophy in general. In terms of literature I naturally looked at the metaphysical 17th Century poets such as John Donne (1572-1631). I realised that an open-minded metaphysical approach (without 17th C. conceits) may be the best solution to my dilemma. Perhaps I should have gone back to the Greeks?

I feel compelled to look at things from a different angle in order to try and explain their significance. I very much like the concept of metaphysics in general and find that it neatly fits my speculative way of looking at life. Indeed this project is full of questions. Metaphysics, as I understand it, is therefore where I might just feel most intellectually comfortable. Metaphysics at its most basic asks two questions as follows:-

What is there?

What is it like?

I could add a further *Meta*-Metaphysical question: Why? - but that would only be complicating the issue. See Appendix: Defining 'God'.

There were three great Greek philosophers who were pupils of each other in turn as follows:-

Socrates (470-399BC)

Plato (427-347BC)

Aristotle (384-322BC)

Aristotle, that greatest of all thinkers, believed that everything has a natural function. He proposed that the natural function of humankind was to reason. He also suggested that reason must be in accordance with ethics. So I suppose, apart from his incongruous defence of slavery: here is someone we can readily identify with in terms of morality, metaphysics and knowledge of what is best suited to the survival of the human condition.

I can, therefore, readily empathise with Aristotle's three main *metaphysical* studies as follows:-

Ontology is the study of being and existence.

Natural Theology is the free-thinking study of god and gods.

Universal Science is the study of logic and scientific first principles.

We can research areas falling between religion and science by exploring the rich fields of philosophy, psychology, sociology and history for example. Many of these disciplines are encompassed by literature in general and this is where most of my influences lie. I have a particular fondness for existential authors such as Fyodor Dostoevsky (1821-1881) and essentially humanitarian writers like Charles Dickens (1812-1870).

Existentialism proposes that each individual - not society or religion - is solely responsible for giving meaning to life and living. So you can readily see where I am coming from. Perhaps I have found the label I am looking for - well for the time being at least. On the other hand if I am too specific I will never come to terms with the wider issues thrown up by a universal, ambitious and wide ranging project such as this. All these designations are fluid and subject to revision as the years go by. Dostoevsky of course constantly changed his mind and eventually went back to his Russian Orthodox religion - as so many do as they grow older.

I find myself in sympathy with the existentialists as mentioned. These may not be fashionable but are nonetheless valid to our arguments. In spite of this I find the rationalism of Immanuel Kant (1724-1804) also worthy of mention. Inevitably further free-thinkers and philosophers such as: Rene Descartes (1596-1650), Thomas Paine (1737-1808), Karl Marx (1818-1883) and Jean-Paul Sartre (1905-1980), have relevance to our theories.

Meanwhile the great scientists and thinkers: Sir Isaac Newton (1642-1727), Charles Darwin (1809-1882), Sigmund Freud (1856-1939) and Albert Einstein (1879-1955) must be acknowledged for the laws by which we understand the scientific concepts of life and being.

I am a great believer in thinking for myself of course but who wouldn't recognize our great debt to these geniuses? You may also find a hint of acquired '*humanism*' pervading my text. I am not sure where that came from. I was never what might be called a '*Renaissance Man*'.

Humanism, as opposed to humanity, is quite an ambiguous word and has, perhaps unfairly, become too much associated with secularism. I think humanism is simply about human-nature. I even think that humanism can be found within the more accessible human-based religions. It may be heresy to say so; but I see Jesus, for example, as both spiritual and humanist at the same time.

Many parables such as our old favourite; *feeding the five thousand* would seem to bear out this theory. I would hope so anyway. However, as this is not a religious text and in view of the truly vast body of religious literature available, as opposed to minimal survival literature, I have not done very much in the way of theological research on this occasion. I am also very aware that you inevitably have your own views in these matters.

I apologise if this particular chapter is judged to be too personal, opinionated, philosophical or irritating. I would crave your forbearance safe in the knowledge that you too have your own views. However an impartial metaphysical approach seems to encourage the essential free thinking needed for an investigation of this nature.

Just think: a different way of looking at life, free from centuries of dogma, narrow-mindedness and mind-control. What an exciting thought. Yet even simpler concepts of philosophy and free-thinking seem to have eluded many of us in favour of those who have controlled our destinies.

I know that innate human conservatism has probably been essential for the practicalities of group survival. However, the actual study of survival itself requires an altogether more inquiring and free-thinking approach. We are all entitled to freedom of thought. To think otherwise is to deny our individual integrity, imagination and inborn intelligence.

Metaphysics is undoubtedly wide open to false accusations of being anti-religious, abstract and unrealistic. It is also very susceptible to hijack by eccentrics and the word is often misconstrued. Yet metaphysics asks some simple, positive and logical questions about humanity in general and promotes some equally logical suppositions about some age old questions.

118

Why are alternative ways of looking at life looked on with such suspicion? Why are we such a conservative species? Why do there seem to be so few truly free-thinkers?

Yet, ironically, I am not too sure that free-thinking is always too desirable for society in general! It is a kind of intellectual anarchy: mercifully free from academic rigour but often vague and woolly. I feel you should honestly and sincerely do your own free-thinking; make your notes and then see what best backs up your conclusions - something to hang your hat on and gain some credibility for your ideas. Then I would say that wouldn't I?

I hover between serious doubts about ancient religion on the one hand and the more mathematical certainties of modern science on the other. Whilst leaning more towards the scientific ways of looking at life I can still hopefully look at things from an entirely different perspective. Thus I have naturally and instinctively gravitated towards metaphysics in my quest for some insight about human survival.

I may not have the answers but I can enjoy the questions; if only to rejoice in glorious freedom from unproven dogma. If we can share this way of looking at these issues then all to the good. Surely metaphysics is all about free-thinking?

Science does not know its debt to imagination.

Ralph Waldo Emerson

Surprisingly this chapter will discuss life-science, often by direct comparison with religion, due to their common objectives of understanding the universe. The science part will look at some surprising human aspects of both physics and chemistry. We may call this life-science for our immediate purposes. However, as human biology and psychology in general are highly specialised scientific subjects; they are beyond the scope of the writer to cover in any great depth.

On the other hand what could be more intriguing than seeing our human being in terms of what we are made of to ensure our survival over aeons of time? Our bodies are the epitome of sophisticated evolutionary living, breathing and thinking matter.

Science is about matter/energy. Religion is about mind. Or maybe religion too is about some kind of energy? What relevance does mind over matter have to human survival? Is religion all in the mind or a mere unproven figment of the human imagination? Yet the brain and the mind are very real things vital to human development - therein lyeth the soul.

Without our human minds, souls or egos we would be a lower type of creature with no science, no religion and no imagination. The mind is where the human spirit lives, the spirit of life itself, the place of survival. Perhaps religion and science are one and the same.

Whether we worship this spirit of life or something even greater remains our own free choice. Does any of this really matter? I would suggest a great deal. Science relies on provable formulae relative to matter, energy and life on earth. Religion relies mainly on faith, history, ethics and morality: yet it too must be influenced by universal energy?

Both science and religion are very important to human survival, one for a better physical life, the other for a better mystical life. Thus both mind and matter are mutually crucial to the continuity of human life.

The one being complimentary to the other in everything we do as human beings. Yet surprisingly both vital subjects are anathema to many.

The following chapters will explore these disparities in some depth due to their mutually strong association with the preservation of our lives on earth. I would ask you to bear with me regarding some of the inevitably more obtuse arguments coming your way in the next two chapters. Science is mainly about facts. Religion is about less tangible spiritual values. Both are fundamental to humanity. Both seek the truths behind our very being. Neither can be ignored in a treatise of this nature.

The history of human faith has existed since time immemorial. It is believed to form a huge psychological part of the human survival impulse as proven by the strong feeling for and against which it still inspires.

Both science and religion are mutually fascinating in intellectual and spiritual terms. Both have been extensively used to control mankind. Do some of us reject science and/or religion from a desire to go it alone? Do we wish to be free from repressive and overbearing philosophies? Free to think for ourselves?

Science has transformed human life beyond recognition in the last 300 years. Religion relies on an ancient and esoteric belief in a better world for humanity and the fundamental power for good as generated by a universal creator. This makes this subject potentially controversial yet nonetheless fascinating for those with open enough minds free from dogma and prejudice.

Even in the face of science; religion still holds strong loyalty amongst its millions of believers. Science and religion have both been critical for the continuation of life on earth in their own ways. The former has achieved life saving potential in a practical sense, the latter in a less tangible controlling, ethical and spiritual sense.

When used for the good of mankind both science and religion have, in general, resulted in the relatively civilised world we now inhabit. On the other hand; when used to divide and kill our fellow human beings both science and religion have a lot to answer for.

Science and religion often sit uncomfortably with each other of course. Each believes that it has the answers to the great questions. Some manage to reconcile their beliefs in both: thus illustrating the complementary power of these two major human philosophies. There are, of course, many contradictions between them.

Due to numerous dynamic discoveries and inventions; science has largely overtaken religion as a credible explanation for the evolution of life on earth. Yet, in spite of many past failings, maintenance of decent ethical values and humanity in general remains critical to both.

Mankind's intuitive belief in God has long been an important and credible part of the survival situation. Never, ever forget the hope, salvation, compassion and fulfilment that religion has brought to millions. Never forget the medical healing capabilities of science.

These age-old spiritual beliefs are difficult to reconcile with science-based convictions about simply living-on in passed-on genes. But before genes were discovered what else could one believe? As a free-thinking modern liberal idealist I am just as likely to be wrong as I am right. Because I am speculating on these matters does not mean that I know the answers. Who does?

Regardless of personal viewpoints these deeply held spiritual beliefs must be respected for their essential humanity. These are entirely natural and hopefully good values of enormous benefit to our real or imaginary survival. I am merely speculating about the actual or symbolic survival of our souls when our lives come to an end.

Like all humans I am simply a product of my time and am certainly not devoid of spirituality or compassion for my fellow man. I believe we live in a phenomenal place for which we can only be profoundly grateful. This is a very moving thought. We have all the means at our disposal to enable our phenomenal survival in this wonderful place we all live in.

What of the air we breathe? What of the pure fresh water of life? What of the lovely green trees, plants and foodstuffs essential to all other life? What of the beauty of it all? What of all those amazing minerals?

Was 'God' the supposed creator a great chemist? Between you and me, he even provided the ingredients for remarkable substances, such as pharmaceutical products, to enhance our very lives. What of the electricity that lights up our lives? It is all there - if we only know how to find it. And miraculous it all is too. No wonder sincere belief in an essentially benign and all-providing 'God' is still so prevalent.

What if for purposes of this discussion on survival we equate 'God' to science for the sake of argument? How did 'God/Science' make such provision for all that we need for survival and even for super survival.

What are the benefits of both organic and inorganic substances to our survival? Where did all the minerals, elements and materials necessary for our survival originate? We even carry traces of some of these essential minerals in our bodies. These materials were essential to make the tools for our survival. Tools to hunt with, build homes, till the ground, make boats and even make weapons.

There are metals, fuels, chemicals and biological resources. All of these are made from the order of 118 known elements consisting of various combinations of atoms. Is the Periodic Table of the Elements mentioned in any modern religious studies? I see no reason why not - this is the Genesis of all matter - life included. I have avoided too much in the way of cross references - but it is strongly recommended that we should all try to acquire at least a basic understanding of this extraordinary elemental table which is easily available. Therein lie the foundations of being: a wondrous God-given formula for life.

Does 'God' in fact live in the atoms - the essence of survival? These atoms: like the universe as a whole, simplistically consist of spinning electrons, protons and neutrons which are the building blocks of matter. These all obey the laws of physics, or the laws of 'God', if you prefer. It could be argued that all life and even intellectual thoughts on religion, or anything else for that matter are all forms of energy subject to all the rules of science. But who can analyse the human soul?

Some of the following paragraphs may require very careful reading and could slightly challenge our grey-matter but we have come this far so please be patient. If you think reading this is difficult then you should try writing about it. The author has no doubt expended a few more brain cells in the process. It is all in the mind of course.

Even if it may all be in the mind, nevertheless, scientific, evolutionary, biological and electro-magnetic forces are all at work in our phenomenal brains. These are significant forms of energy. It is said that when we eat fish it is good for the brain, or carrots for the eyes - is this literally food for thought? The fact is that there is an infallible law which states that chemical matter and the kinetic energy, or motion derived from it, can neither be created nor destroyed.

It could be argued that life is essentially a form of motion, or kinetic energy. Total energy, of which life is just one example, must therefore remain constant over time. Remember the formula that Survival = Life / Time? Thus survival becomes as fundamental as the laws of physics once life had duly evolved from other forms of energy. To put it more prosaically: Life is a miracle, Energy is undeniable and Time is eternal; so Survival is everything.

As we move on to discuss the law of conservation of energy we will soon grasp the idea that conservation of life (i.e. survival) is a fundamental example of this truth. Even if life became extinct the compound mass or energy of all organic, living or dead, creatures would be transformed into alternative forms of energy such as coal or oil.

Equally, logically but no less amazingly: life evolved from other forms of energy such as heat, gas, carbon and organic matter, rather than being truly created as such. In lieu of these scientific explanations the miracle of life could only have been thought of as heaven-sent by our ancestors who obviously knew no better.

No wonder 'God' as a 'Creator' was envisaged before a true understanding of these complex scientific formulae evolved. This is not to disparage the achievements, sincerity and good will of our ancestors. One of the most basic laws of science is the Law of the Conservation of Energy as already stated. Back to $E = mc^2$ - this states that matter, or energy, can neither be created nor destroyed. Now that is interesting. No creation? No destruction?

When matter appears to be under destruction it is of course being converted to alternative volatile kinetic energies such as heat/gas. When we convert wheat into bread we are ultimately converting organic matter into the alternative energy of human existence and intelligence. When we drink wine we do the same. No wonder they have such symbolic significance to Christian believers.

Survival must underlie much religious doctrine as confirmed by the many parables about bread, wine, fish and so on - even ideas of controlling the consumption, or prohibition, of certain types of meats, such as pork, is often of practical survival value. Some foodstuffs are less wholesome than others in a hot climate. Also fasting and limiting the intake of food in general are very fundamental lessons in preserving and rationing rare supplies of much needed protein. We are what we eat and food is essentially fuel to create living energy for our human well being.

We are surrounded by useful materials and edible substances to be promulgated into life itself. So where did all this matter and energy come from and where is it going?

The next question, which will never, ever be answered, is WHY? How can we honestly even begin to understand all this? However, matter can be almost infinitely adapted to the benefit of all concerned. The chemical balance has literally achieved our body and soul. No wonder we worship and seek to preserve our very being. No wonder we question the very miracle of life. No wonder we find it difficult to equate science with theology.

Our lives are ultimately based on scientific principles. These are the essence of survival. As a matter of interest there can be as many as 60 chemical elements contained within the human body. However, there are 12 empirical human elements in proportions as follows:-

H 15,750 N 310 0 6,500 C 2,250 Ca 63 P 48 K 15 S 15 Na 10 Cu l6 Mg 3 Fe 1

Hydrogen = H, Nitrogen = N, Oxygen = O, Carbon = C, Calcium = Ca, Phosphorous = P, Potassium = K, Sulphur = S, Sodium =Na, Copper = Cu, Magnesium = Mg and Iron = Fe.

Roughly 96% of the mass of the human body is made up of just four elements: oxygen (O), carbon (C), hydrogen (H) and nitrogen (N). We will note that most of this is water (80% H^2O) in terms of H (hydrogen) and O (oxygen).

So not only were our remote origins actually in water but this essential substance is still available and plentiful enough to enable our continuing survival. Water is the essence of our beings. Water is crucial for survival. If there is one resource we should be conserving, above anything else; then fresh clean water is like liquid gold in terms of survival. So think about this next time it rains.

Strange to think we are just as much a product of rain as any other land based mammal or plant. We need never see rain as a problem (providing it comes at regular times and in reasonable quantities that is). Is it any wonder we used to worship sources of life-saving water? Even the bitter salty mineral rich sea is teeming with life at our disposal.

Our reverence for water extends from holy wells to holy water and baptismal fonts. All our life forces are subject to many natural materials of course. Carbon too is a vital ingredient. We each inhale about 11,000 litres of air per day: of which about 20% is oxygen. Of this about 5% by volume is exhaled as carbon dioxide (CO_2). This together with industrial emission of this ubiquitous gas leads to a considerable output when multiplied by the estimated 7.0 billion humans surviving on the planet at the present time.

This is a good example of matter/energy being converted and recycled rather than being destroyed. Luckily we live in a very big place indeed and most carbon dioxide is naturally recycled in the oceans and by plants which help convert it back to oxygen again. The other vital minerals and chemicals in our bodies are presumably largely supplied by what we eat. Everything we eat comes either from the mineral rich ground or sea.

Our inherited spiritual energy, life-force and divine heritage, i.e. our very lives or souls, will form a large part of this discussion. These have been the subject of scientific and theological speculation over many centuries.

Insofar as our essential life-forces are concerned there are additional chemical and electro-magnetic energies at work. In nature there are electromagnetic fields ranging through a spectrum of varying wave lengths. There are 7 main 'classes' of electromagnetic radiation ranging from short wavelength gamma rays, X-rays, through ultraviolet, visible and infrared light, microwaves and finally long wavelength radio waves.

No wonder mankind believed in magic before the discovery of these phenomenal forces. And what of gravity, earthquakes, storms, seasons, tides, time, night, day, heat, cold, etc before we even begin to understand electricity and magnetism? What in fact are the essential scientific forces of life itself? There are certainly some complex biological, genetic, chemical and electro-magnetic forces at work in our bodies.

There are three main aspects of the human personality known as the id, ego and super-ego. If the human concept of 'God' does in fact largely lie somewhere deep within the Freudian id, ego, or super-ego (or soul if you like) as the writer suspects; then maybe we have found some kind of believable and plausible explanation for 'God' as he lives within each of us.

This is in fact our Life-force - life of course being a mere fragment of the bigger time/life/matter/energy/space picture out there. This is merely the more understandable human manifestation of something far greater and far less tangible.

Think about this if you will. If we consider that the human personality is divided into three main compartments: the ego is the best known part of our personality. What is the ego anyway? Perhaps that is the rational part. Then what of the id, and the super-ego? The id is the unconscious foundation of the intellect. Certainly the super-ego is known as the receptacle for consciousness.

The super-consciousness of the super-ego is important to my argument. The super-ego is the receptacle of all that we have consciously learned since birth. It takes the place of the Oedipus complex. It is the powerhouse of the human mind. It is possibly where we know right from wrong.

Links with the cultural super-ego, instinct, ethics and religion bring us closer to some understanding of this complex subject. Could the notion of 'God' lie predominantly somewhere in the super-ego part of the psyche? This is where we sort it all out. This fits in very well with the hierarchical nature of mankind. Everything is stacked in order. This is the filing cabinet of all we know. The status maybe lurks here too. This is where our all powerful imagination comes into play.

Thus God, by whatever means He is perceived or denied, has a natural place in the ordered mind. Thus God is maybe part of a certain logic inherited from our many forbears? Creating Gods certainly confirms our human obsession with status.

On the other hand God is perhaps more a product of ancestral memory, or human instinct, rather than of rational thinking? So possibly He is also part of the unconscious id rather than of the rational ego or of the conscious super-ego? Ask your shrink.

Perhaps God progresses through all compartments of our entire personalities during our formative years before either being rejected during adolescence or finally coming to rest somewhere in the super-ego? In other words we make a conscious decision whether to reject, doubt or accept God but an instinctive unconscious belief is passed on to succeeding generations to be reprocessed in due course.

The germ remains. Some of us never reach any conclusions in our entire lives. So the endless cycle of belief, semi-belief and non-belief goes on down the centuries according to prevailing cultural norms and our instincts.

The intellectual idea of God therefore is maybe Life itself as expressed within the personality? I am increasingly persuading myself that 'God' may be Energy itself. We live therefore we are. We exist - if we did not exist then we would not worship or believe. If we are dead our Life-force is no longer valid.

This is where most religions differ from science. Most religions believe in an unproven so called after-life. This is not based on any logical modern ideas of simply living-on in the genes of our children but on unproven ancient concepts of heaven, hell, etc. As a survivalist it is fairly obvious that I would take the former view. I expect that many scientists believe we die and then live on in our genetic offspring. Now that genes have been proven this puts a whole new emphasis on the subject.

I am suggesting that our traditional concept of an understandable human God is most readily expressed within the personality - or the mind if you prefer. This is more a theory than an explanation of what He might well be in the wider and more universal conception of God. In any case this is considered to be far too complex for true human understanding. However, as already suggested, we may safely presume that the human concept of God may in fact be some form of universal kinetic energy of which life on earth is a chance by-product.

Remember we are imbued with ideas of fantasy from an early age and these could well be related to unproven religious suppositions over many centuries; when we did not know any better. In truth it is proposed that we actively worship, or express, him, simply by surviving to live and living to survive as stated in the Introduction. This is an entirely natural explanation for something which is otherwise supernatural.

It is even feasible that mass religious fervour is some kind of instinctive accumulation of human electro-magnetic force where the needs of the congregation at large takes over from those of the individual in the manner of the well know beehive or mass migrations and so on -

like a kind of mass hypnosis which relies on the willingness of the believers to give themselves over.

This is a bit like mass-hysteria, accumulated passion or a shared emotional outpouring. We have seen similar examples of mass human emotion and hysteria at the funerals of much loved and over-idealised public figures for example. This is an expression of pure living energy and energy of course is a God-like commodity. Thus humans are capable of creating a mutual sense of God to fulfil a common spiritual need.

To this day Man's hunger for God still has a lot to do with the remaining mysteries of life on earth as part of the bigger picture of an unresolved and infinite universal energy beyond comprehension. Remember that early humankind simply did not know what energy was.

What else could they do but worship the majesty and beauty of this great so-called spirit? They could only guess that an even greater person, or creator, was responsible. Their great kings, prophets or interpreters (through those who could write about these mysteries) then assumed huge holy stature and gained an infallible superhuman spirituality to their followers.

We now have an idea what energy is. We have put it to good use but will never understand its ultimate purpose. Before the advent of science these mysterious energy forces were thought of as magical, supernatural, mysterious and mystical. This is vital to our argument.

Like the air or the 'firmament'; energy would also have been associated with the creation, spirits or even the Holy-Ghost. We may now think of these as Life-energy or the Life-force. This is still very hard to define in scientific terms. Therefore the Life-force (because force it must be) still assumes huge mystical importance in lieu of convincing scientific analysis. These Life-forces clearly originate from the Astral energy of the great trinity of sun, moon and earth of course. Biological Life-energy is clearly associated with heat, light, electromagnetism, chemicals, organic matter and proteins over time. But what is it?

These vast energy resources remain magical and mysterious and have always had God-like significance. Therefore we can safely attribute most life-giving kinetic energy on earth to the sun which has been worshipped since time immemorial. I use the word kinetic advisedly to relate to the perpetual motion of life. The warmth of the sun has clear daily associations with time and life itself. Gravity, climate and air pressure too have their place.

The Sun-god has quite logically been perceived as both energy and God at the same time. Indeed the sun must have seemed the epitome of survival to our forebears. Nothing could be more natural. The life giving energy of the sun is a kind of electro-kinetic energy if we think about it.

By turning to science again we may certainly look at potential and kinetic energy as being at the source of the Life-force. This idea of the importance of motion and kinetic energy, together with life, space and time as divine entities, must also have scientific explanation. A tentative conclusion is that science and religion have some parallels in terms of mankind's incessant urge to survive - the scientific view is that parallel lines will never meet. This idea of parallel lines makes us think of infinity; so one wonders just how does religion view this concept.

To summarise, in spite of the energy of the great universe, perhaps a large part of our conception of a man-centred human God actually lies deep within each and every one of us. This is the supposed religious-gene or soul referred to elsewhere.

When multiplied by the souls of all the individual members of a religious congregation, worshipping together for example, then a powerful cumulative and composite chemical and/or magnetic energy (aura) is generated resembling that of a perceived all-encompassing 'God'? The Welsh know this highly emotional state by the almost untranslatable word:-

Hwyll.

You can sense this tangible sacred magnetic energy when witnessing mass acts of worship.

Fascinating books about the powerful human 'aura', life and vital energies (together with 'Prana', 'Chi', Astral and Spiritual Life-forces) are to be discovered on the Mind, Body and Spirit, Popular Science and other Metaphysical bookshelves. Further research on these fascinating energy sources is strongly recommended.

Or is this residual, sacrosanct energy from somewhere deep in our ancestral memory as already discussed at some length? The true nature of what this latent energy is has yet to be determined and what exactly is memory; or religion for that matter - some kind of imprint on the brain? Who says religion and science are not interrelated?

14   RELIGION

And God said, Let us make man in our image, after our likeness: and let them have dominion over the fish of the sea, and over the fowl of the air, and over the cattle, and over all the earth, and over every creeping thing that creepeth upon the earth.

Genesis 1:26

Whether we worship 'God', fundamental energy, creation, life, the sun or even the eternal constant of time, is down to our own faith and culture. This instinct for worship motivates us to wish for a better world. Religion in particular is believed to have enormous relevance to human survival. It obviously relates to traditional codes of human morality, spirituality and the quest for better things.

Whilst the principles of human faith have been approached from an impartial and detached humanist view point: it is hoped that the writer has treated this emotive subject with all due respect as one of the major foundations of ethics, civilization, humanity and survival in general. Equally, science is considered to be a further positive aspect of human survival; in spite of its potentially harmful character in terms of the horrendous weaponry used on the innocent and vulnerable.

Many of us now survive due to the benefits of modern medicines and technology. Human survival is also influenced by genetics and this subject makes an inevitable appearance during the course of this exposition so far as it can be understood that is: genetics as opposed to Genesis.

I find it impossible to understand the many theological connotations of universe, energy space, life, time, nature and creation in general. I can only assume that worship of God relates in some way to these inconceivably vast repositories of cosmic energy and spiritual being.

Perhaps Energy is another name for God? Could God, in the widest meaning of the word, be pure Energy? Energy spiritualised and humanised by the mind. See Appendix: Defining 'God'.

I believe that human concepts of God are largely a product of the human mind, ego, imagination and soul. I think that worship is related to the natural instinct for Survival in terms of Life/Time and hope for a better place. I see this worship of life in particular as entirely natural, genetic and instinctive. Yet who knows 'why'?

The existence, or not, of God is a metaphysical question for each and every individual to determine for themselves. The inalienable humanitarian right of freedom to worship, or not worship as the case may be, has been essential to these studies of human survival. One of my favourite quotations is one made by the surprisingly metaphysical St Anselm (1033-1109) as follows:

God exists because the concept of God exists.

The presence of one type of supposed deity does not preclude the presence of any others of course. Before the creation of one God, known as monotheism, they used to have one for every conceivable occasion. This is known as polytheism and forms part of the great Hindu religion.

This Hindu belief in many subsidiary Gods under the authority of one Supreme Being, known as Brahman, allows for some flexibility and has a certain infallible logic. They also refer to the Trimurti as follows:-

(i) Brahma the creator

(ii) Vishnu the preserver

(iii) Shiva the destroyer

By contrast, the essentially western mystical concept of the Holy Trinity of God the Father, the Son and the Holy Ghost, lies at the very heart of the mystery of the Christian faith and is well beyond the scope of this writer to attempt any kind of explanation.

The relatively simple concept of a Good man known as Jesus Christ still remains a powerfully humanistic (for want of a better or more appropriate word) and attractive aspect of the Christian faith: in all its manifestations.

Perhaps some of the mysteries of our western God are caused by some confusion between the two extreme versions which we are addressing here - the outer and inner ones - the universe versus the soul. It could even be thought of as God being the essence of the universe and Jesus being the essence of the soul? In other words; is God simply a figment of the human brain on the one hand or something far, far bigger on the other? Or is he both?

Perhaps the best bet is to concentrate on the more easily reached inner human God, equated to Jesus, rather than the more inaccessible outer superhuman universal one for the sake of this discussion. I think many Christians do this. Similarly we can see Mohammed as the messenger of Allah in the Muslim tradition. Moses or David may also be regarded as major human prophets in the Jewish tradition.

Any ideas of mine about religion in general are a small and insignificant part of a vast literature on the subject. However, most religious literature is either propagandist or predisposed towards belief. Some of it, at its most extreme, is hysterical, fixated and fanatical. Yet all this tells me is that there is a fervent natural psychological need for spiritual understanding. In fact religion is an age old obsession which remains both strong and compelling for huge numbers of people around the world. These religions are usually fundamental to the prevailing culture, laws and moral values and, as such, are of crucial significance.

Man has always hungered for God. Much of religion quietly fights shy of science and the proven mathematical formulae that go with it. At best there is an uneasy truce between them. However some are able to reconcile both science and religion quite readily. Yet there are obvious parallels. It is not a case of one or the other for many: thus proving how important both these disciplines are.

Apart from lonely ascetics, many religious cults rely on both religious and communal activity of mutual benefit. In fact these cults give us a rich example of human survival at its best and most rewarding. They are often closed to outsiders but certainly indicate the importance of 'family' as already discussed.

Even in highly advanced cultures, such as the United States, religious dogma remains powerful and universal. Little of it however directly relates to survival. Some sects, in particular, address the evangelical spiritual needs of the poor and needy who derive great satisfaction from this vital outlet for sorrow and joy. Some religions regrettably allow little of the essential open-mindedness necessary for the study of survival. Is this why the subject is so little studied?

The discipline of religion and the anti-discipline of survivalist free-thinking make uneasy bedfellows. The underlying discipline of religion in general fulfils many needs but does not always encourage open-minded discussion about survival in general. Pity really.

God as an ethereal, supernatural, intangible and mystical entity is often feared, misconstrued and misunderstood. The devil is considered to be the evil antithesis of God. So in order to promote a more human, kindly and accessible version of God a Messiah or sanctified human representative on earth is a popular way of making God and man one and the same.

This almost subliminally suggests that if God and man are in fact one and the same: then he actually lives in us all. There is simply an identifiable human role model for us all to emulate if we possibly can. See remarks about an inner human God perhaps living in the soul of each of us.

Even the ungodly can identify with this idea in lieu of a better explanation. It is really about the irrefutable Life-force living in us all. In some religions the concept of human godliness is taken even further in the adoration of various humanised saints, prophets and so on. The 12th C. cult of the Virgin Mary in particular has remained extremely popular.

Three revered human representatives of the major Abrahamic religions, such as Islam, Judaism and Christianity, would be Mohammed, David and Jesus for example. In the Jewish and Christian faith these are sometimes known as Messiahs (anointed ones) and are regarded as Messengers or Prophets. There are many more human prophets, leaders and mystic representatives of the major religions of the world beyond the scope of this essay.

So religion has a very long and complex history. Can you imagine a world without it? Who am I to know? So let us look again at science to see if some further answers can be found.

My personal belief is that these questions relating to the concept of an infinite universe will never, ever be resolved. Even though we can see a small portion of the universe; the infinite vastness of it all eludes comprehension. The key word; infinite, implies an unknown entity without beginning or end.

How does science deal with such ethereal and un-provable concepts as a beginning or an ending? Science by its nature must see the infinite very differently than those who are religious. Does science have an explanation, law or theory of just what constitutes the infinite or beginning or end for that matter?

There is an essential difference here. If the beginning equates to creation and the ending equates to destruction then where does all that matter come from or go to? Why does Matter 'matter' anyway? What is it? Answers on a stamped addressed envelope please. First prize a trip into infinite space to find out. Second prize a trip back to explain it all. If you live so long that is.

To ask a hypothetical question from an unbiased perspective: which is the true faith: mine or yours? Why are there so many supposedly infallible religions? It would be naive to suppose that man-made religions are all as perfect as they would have us believe. They are the product of an all too imperfect and divisive human condition as already discussed. Some, but not all, are imbued with all the snobbery and class divisions reflecting their very human rather than divine failings.

Remember our human propensities to competition as opposed to co-operation? Peace versus war. Love against hate. Good and bad, greed versus sharing. The list of our contrary transgressions and hypocrisies does little to make us as perfect as we might like to think we are. Most western religions make allowance for this of course. A weekly top-up cycle of worship and cleansing is deemed adequate to at least keep us on track. This, together with a day off work, does much to promote our belief in perhaps a better life out there.

Compromise is never all that far away in these matters. We are all human beings after all. I suppose they thrive on the idea that we may at least try and be better for the time being. Sainthood comes later. However, despite their patent imperfections and divisions these religious foundations must surely be fundamentally on the side of right.

The all too human religious formula seems to have worked well enough for centuries and continues to fulfil the spiritual needs of millions. To deny this is to deny something deep in the ancestral memory. There must surely be a religious gene forming part of our survival tool kit? Why else would it have continued throughout history? These things run very deep whether we realise it or not. Remember the emphasis I placed on human caring? Yet religion in general has been written of as one of the greatest delusions in the history of mankind.

Despite our cynicism it is surprising how we might desperately turn again to our latent religious origins lying deeply, instinctively and subconsciously in our personalities for when the need arises. When we can no longer cope we need something to fall back upon. We need to hand responsibility on to a greater authority. We need a saviour whoever we are.

We are only human after all. What could be more natural than that? Even when it seems illogical, or hypocritical; there is no getting away from centuries of inborn ancestral convictions. There must be no city, town or village in the world that does not have a holy site, house of prayer or sanctuary to this day. We should stop and seriously consider why. Man's craving, weakness or primitive hunger for God still lives on.

Can you imagine any civilised community without these icons of past beliefs? Heritage is sacred to many and long may it remain so, as long as we respect our ancestors; without whom we would not exist. Whether they were always right or wrong will remain a subject for further debate. How right are we now? How right will we ever be?

Yet religion may not always be a particularly rational thing to do. It is particularly difficult to wholeheartedly believe in unless you were born to it or have become born-again. In many countries it undoubtedly belongs largely to a bygone and perhaps less sophisticated era. In other countries religion remains a fundamental part of daily life of huge value to all concerned.

This is not to deny the huge body of positive good lying outside the various faiths either. Individuals quietly getting on with helping one another: free from the ritual, pomp and ceremony of organised religion. Some even find this emphasis on power, control, ceremony and hierarchy contrary to the basic human values considered so vital for survival. Yet many of the poor, simple and ordinary people around the world simply love the allure and glitter of the rich and powerful as epitomised by some branches of the Christian Church for example.

Yet, let us never forget the truly humanitarian work done by many religions in terms of caring for the spiritual and physical needs of the poor, dying and dispossessed. The writer has visited these dedicated true Christians in the African bush and there is no praise high enough for them. This is just what Jesus would have done.

There are even some saint-like members of these faiths and we can do nothing but admire their commitment to their fellow man. They are a more important and vital part of the survival conundrum that they might at first appear. This is why this writer believes, despite some personal misgivings about many aspects of their formal constitutions as mentioned above, that they are essentially on the side of good and on balance should be acknowledged as such.

In the opinion of the writer these alone more than justify their existence and do much to counterbalance any perceived all too human excesses which they may exhibit from time to time. They have both the best and the more questionable sides of the human condition and are no different from any other human organisation. We are all too human when all is said and done. My only wish is to see them continue to change and adapt to modern needs or they will inevitably continue to lose the faith of many and that would be a great loss to their vital role in promoting good for the ultimate survival of mankind.

Probing these concepts is quite cathartic. Just because the writer sees things from various, perhaps radical, viewpoints is no reflection on the integrity of his viewpoint. Well the example of the extremely radical Jesus is no bad example to the rest of us. Unfortunately when religion is discussed, in any way, shape or form, opinions too readily become hardened either for or against. In fact to slightly misquote a Shakespearian gem from the ultimate *non*-survival play: Hamlet:-

Methinks the lady doth protest too much

In fact why does religion matter so much to both believers and non-believers? Religion should never be a contentious issue. It should be looked at objectively, dispassionately, sympathetically and with an open mind. Remember there are very many versions of how to interpret the holy teachings. There are many religions to choose from. There are even more interpretational divisions amongst each religion in turn.

However, in terms of survival; religion must be seen as a basic human spiritual need. It has existed in one form or another for hundreds of years and must always be afforded due respect. What one believes in, or not, as the case may be; is not our concern here. Some question, other are in denial. Some are not sure. Not many are not influenced, if only slightly, one way or the other. This is why it is featured so prominently here.

Almost everyone has at least some pseudo-religious ethical values. These often originate in the prevailing culture and simply instinctively knowing what is right. These are the best kind of lingering spin-off from the best of long cherished religious values. These personal ethical values seem to be an entirely natural survival strategy. Thus the world continues.

Those who implicitly believe are very much to be envied. Religion is an important human issue; plain and simple. It inescapably relates to both real and spiritual survival. It always has and always will for as long as it survives and we would be significantly worse off without it.

We need *good* above all else for survival. Thus the world goes on. It is the best we can do I'm afraid. Many are comforted by the otherworldly ritual of prayer and worship and it would be extremely churlish to deny this. To them God is very real and therefore cannot be denied whatever you or I might think. To deny this is to deny common humanity and humanity is vital to our survival - it always has been. This does not make the writer an adherent, or even an apologist, for any specific religion. This is more about respect for all legitimate faiths.

These generally revered and respected ancient religions, together with the evolution of science, must have played a very significant role in the structure of society, as it has become.

Mankind seems always to have had some form of spiritual belief, no matter how imperfect, to help to explain the unexplainable. Apart from the obvious controlling aspects of religion it would be only natural for many to worship this deity and for others, of a perhaps more open mind, to question its validity. It is often said that we have an inborn religious gene. Much of it may well be found deep in our subconscious ancestral memory.

The sense of *déjà-vu* in sacred places can be remarkable. Did some of our countless ancestors worship here or in a place just like this? Surely we all recognise these attempts to bottle-up divinity as it was perceived at the time in these wonderfully sacred old buildings?

140

Or was the compound and accumulated fervour of their devotions somehow imbued into the ancient masonry? Has the sheer weight of residual spiritual energy somehow enhanced the holy atmosphere and ancient calm of these wonderful places? It is certainly like stepping back in time.

Does this trigger something deep within us? See chapter on Ancestral Memory. These divine impulses could well be experienced when visiting some place of extreme man-made spiritual beauty and sanctity; such as a cathedral, mosque, temple or synagogue.

Some of the greatest architecture, art, music and literature owe their origins to religion. Some religions are so deeply imbued in everyday life and custom that they are fundamental to the survival of whole cultures. They are based on certain logical and historical doctrines. They are a code for living of incomparable consequence to humanity as a whole.

Despite a sometimes dark and all too human underside; most religions have an astonishing beauty. I certainly could not envisage a world without them. Many non-believers would hold similar Pantheist or Theosophist views in the awesome face of nature. Beauty in many forms remains a powerful spiritual element in our lives. Beauty; as opposed to ugliness. Good as opposed to bad.

Some even see beauty in the mathematics which helps explain our universe. Some would see glory in simply everything around them just by the miracle of being alive. Who has not at least once in their lives woken up to the glory of it all? Who has not soaked it all up - the miracle of our very being?

Life can be sweet. No wonder survival is so important to us. If nothing else there is always the beauty of it all. Others might eagerly believe in flying-saucers, ghosts, fairies or other paranormal events. It is often difficult not to be at least mildly superstitious.

Some worship at the secular or intellectual altars of Aristotle, Beethoven, Raphael, Shakespeare, Newton, Darwin or even Paine. Some worship their athletic gods, teams and national heroes. Royalty, celebrities and rock-gods are all alternative candidates for an inexplicable human instinct to worship.

A few are of course extremely sceptical about some of these pseudo-gods. Some even have a strong, ironic and almost godlike fervent belief in their being no 'God' at all. They may have a valid point as there is as little proof of there not being one as there is any rock solid proof that there is one. How much they are in strong denial due to subconscious religious disillusionment is a moot point. Atheists take note.

Many, of course, seek answers to the great questions in science. In spite of the prevalence of science based atheism and agnosticism - in our modern technological age - not many human beings do not retain at least some vestige of ancestral spirituality or superstition in one form or another if they are honest with themselves. This may well be an instinctive human need related to deep-seated unconscious survival instincts. See Introduction.

Some, perhaps with justification, are still using Darwinism to validate their atheism - although most have now moved on from this.

Others, whilst sympathetic to all other reasonably credible faiths, have little personal conviction or allegiances to any one particular faith. They are open to the best of all faiths and philosophies in general but remain agnostic at heart. They are free-thinkers. The writer can strongly identify with this viewpoint. However, this may be something they regret from time to time. This is because people naturally like to be identified with some particular communal values. They are going it alone. We all need a social identity. We all need at least something to believe in.

Without group identity we are in a kind of vacuum. However, they feel this to be the most honest approach. They acknowledge that there is something wonderful out there and also within themselves, as already discussed, but have difficulties in associating this with any one of many deities available to them. I expect this strikes a chord for many in this day and age.

The problem also comes when trying to worship a higher authority, without any validation: seemingly because He is all powerful. This is all about hierarchy rather than humanity and seems to be in fundamental contradiction to many of the basic teachings of Jesus. The writer recalls being extremely dispirited as a boy by the 'hell, fire and brimstone' and predestination aspects of Scottish Presbyterianism at the time.

There are very many contrasts between say the Old Testament Book of Daniel and the New Testament Gospel according to St Luke. These contradictory aspects, of what is ostensibly the same religion, seem to be an unbridgeable chasm. No wonder many of us do not take the Bible literally, or at face value - but ironically see it as simply one of the most stunning works of literature ever produced.

Other inspirational works of faith, such as the Koran, are little known by the writer but are undoubtedly of immense spiritual significance to Islam. This has been a particularly successful religion which pervades every facet of life and can be readily interpreted and understood by all. There are many other beliefs of course and these all demand due respect.

This belief in survival, in one form or another, is so fundamental to our culture as to be collectively accepted, or subliminally so, by most of mankind. Regardless of our faith, or perceived lack of faith, we would be failing this whole marvellous but inexplicable universe if we did not have a deep instinct to survive as a species; no matter what the odds.

We may even be arrogant enough, as the master species, to believe that without the survival of our superior intellect; none of it would exist anyway. Continued existence for each and every life form, by whatever means, is a powerful instinct by any standards. Thus the world goes on. Just witness the miracle of a baby being born if you don't believe me.

Consider also the image of motherhood. This is the essence of survival. The origin of the Christian religion in particular, which has survived for about 2000 years, was even partially based on this premise. The tragedy of a good man dying in agony for his beliefs does not go amiss either. Humanity and civilisation in general are founded in these symbolic images. In lieu of something better we lose this mythology at our peril.

Even though I have attempted some open-minded analysis of the subject the existence, or otherwise, of 'God' is not in question. We surely all have some spiritual, ethical and moral values by whatever name you want to put on them? Principles are probably as good a name as any.

Before we go any further the writer strongly believes that we are an imperfect species by any standards and have great difficulty in attaining the saint-like state of utter perfection we may aspire to. It is the trying which counts. If, like many millions, you sincerely believe in your own version of God, within whatever credible framework you may choose, then God truly exists for you. This is your ideology and must be respected for its incontrovertible value to the survival of your own soul (Life-force) and the souls of your fellow believers.

However, your God must rather be one of caring, sharing and compassion rather than a cruel and unforgiving one to be most intellectually and morally acceptable. Cruel Gods were a product of cruel times. Good Gods were an expression of better and more enlightened times.

Providing religion is based on truly caring for each other; rather than just for oneself at the expense of others; then it can only be of benefit to mankind. Even selfishly caring for the redemption of oneself is probably preferable to not caring at all. These are personal values and are all part of the survival conundrum. This belief in a higher authority is concluded to be one of many fundamental human needs.

I believe we are all spiritual beings in one way or another. Some have yet to find a spiritual home of course. I too am undoubtedly 'spiritual' because I live, breathe and survive as part of a greater but not yet fully explained Energy called Life. However I have yet to reach any really honest personal conclusions because I am yet to be convinced of all the truths of these matters.

Because I am free from any specific spiritual allegiance and am unable to worship an ill-defined and unproven deity I am more inclined to seek sincere, logical and rational explanations in the light of present day scientific knowledge. Therefore, as an uncommitted free-thinker, I am at liberty to look at these matters from an objective, rather than a subjective, viewpoint.

I do not know the answers of course. However, I feel qualified to speculate due to strong associations between science, religion and human survival. Therefore, I feel free to put forward some personal theories about our all too human concept of 'God': which may, or may not, find a little recognition by both science and religion.

See Appendix - Defining 'God'. Many questions of course remain to be answered. Some of these may possibly elude us until the end of time.

15   WORK

No work is insignificant. All labor that uplifts humanity has dignity and importance and should be undertaken with painstaking excellence.

Martin Luther King Jr.

In order to survive in the modern world many of us have to work; if we can get it. In much technology driven western economies there are more people than jobs. Our tools have outgrown us.

We work in order to earn money to feed and house ourselves and families. Work could well be defined as an exchange contract between human beings. We all work in the common interests of society. This society is now on a global scale. Work is now usually corporate, disciplined, formal and institutional for many in the towns and cities. Work for many may be in the financial, service or manufacturing industries. Some are employed by others and some are self-employed.

Work may broadly be seen as manual or administrative. These became known roughly as blue and white collar occupations. However, there are many hands-on professions which combine manual and technical skills such as engineers or surgeons for example.

Management covers a wide range; then there are all the financial, sales and retail professions. Media, publishing and press too are of importance; as are computing and communications in general. There are a number of creative professions and it is even rumoured that some brave souls manage to write for a living.

Education is considered to be one of the most vital of occupations in terms of affording better opportunities for those who otherwise might struggle to survive. Most occupations require various levels of practical, theoretical and academic training. There are multifarious vocational and educational qualifications. So, old-fashioned classifications of skilled, semi-skilled or un-skilled workers are no longer as significant as they once were.

There are primary industries such as mining and secondary industries such as manufacturing, transport and distribution. Due to the huge increases in technology and urban living there are now greater numbers of white collar workers than ever. These have seen a burgeoning middle-class, with corresponding wealth and education for some, and greater divisions between rich and poor.

Whilst much agricultural and livestock work is now mechanised: large parts of the world still rely on traditional small scale peasant and community farming. In essence bartering has now given way to payment in exchange for goods and services. Thus we work for money to spend on survival and quality of life. So we may benefit from a few paragraphs about the money we might earn before looking at the benefits, or otherwise, of working.

Money has long been used as a convenient means of establishing the value of items being traded. Money as a means of trading in vital commodities has now become universally established as a product in its own right and a means to an end rather than the ancient practice of direct exchange or bartering. Thus the so called money-markets have become an artificial commodity which has led to further non-productive and artificial wealth creation of little direct value to the ultimate long term survival of us all.

Recent problems in respect of extreme greed, corruption and rampant capitalism have become obvious. Yet other experimental political systems such as Communism have been found not to work. This is due to the restrictive nature of oppressive and over-disciplined regimes. These used Marxism as justification for the repression of the vast majority in favour of the privileged few. This flew in the face of the relative freedoms and prosperity of the more beneficent 'liberal' capitalism from which the west was undoubtedly flourishing. Well for some of us; according to geographical location at least.

Maybe long overdue improvements to our substructure can result in much needed work for our unemployed. People who create things rather than just trading in them? Most work is creative in one way or another but real creation of real things of benefit to society can be the most rewarding of occupations. When the substructure is failing there must be much to be done to the benefit of all concerned.

Everyone needs to feel they can contribute to and benefit from society by working for reasonable wages so that they can maintain some status and dignity. Creative/meaningful work should make you proud not ashamed. Employment needs to be profitable both to management and workers alike. Failure to respect each other has led to past industrial unrest on a huge scale and benefitted no-one in the long run. Work is very much about dignity, teamwork, and friendship.

It is obvious that early mankind could only have survived by co-operating within the family, tribe and nation by specialising in one way or the other: working together in other words. This implies a hierarchy of seniority where the oldest and perhaps wisest took on teaching, leadership and advisory roles. This was of course the beginnings of a class system based on occupation and a fundamental need for status and belonging.

Everyone needs to feel they belong and are useful to their community in some way. Because of technology and 'over-population' this is now denied to many. In prehistoric times the youngest and most physically fit men in particular would have taken on the hunting, fishing and combat activities and competitive roles in general. Some group members would have served apprenticeships according to their individual creative talents in terms of working with wood, stone and metals (artisans).

These skilled specialists would have led teams of semi-skilled and unskilled associates in group projects such as building of shelters and so on. Other groups of women, in particular, would inevitably have been most associated with domestic skills surrounding food preparation, weaving and bringing up the children - all working for mutual survival.

The traditional role for women has lasted to the present day in some parts of the world. However, the late 18th/19th Century Industrial Revolution, followed by long overdue feminism, saw the role of women in general changing hopefully for the better. Even during the Industrial Revolution, when women worked mainly on factory production lines, they would have often taken time out to rear their children. They would then become totally reliant on the wages of their husbands as so called breadwinners before returning to work in due course.

With the advent of feminism, modern work practices and labour saving devises it was seen that a vast and much needed untapped labour supply in terms of female employment could now emerge. Perhaps for practical, political and economic reasons; increasing use of women in the workplace has resulted in one of the biggest social changes ever known to humankind. In fact many young mothers now have to return to work far sooner than their own mothers did. The long term effects, for good or ill, are yet to be decided. In general many women are thankfully far better-off and independent than they ever were. Many women now hold senior roles of course.

The demise of heavy industry has, in turn, altered the traditional role of men. Many previously blue-collar men in particular have also become adapted to modern technology. The consequences of modern technology in terms of enhanced pay and working conditions has improved but, of course, there are fewer jobs for those who did traditional manual work.

Work and its consequent wealth creation is of course vital for our survival if only for the money it puts back into the system. We may not view our corner of the labour market as particularly altruistic: yet all work should be for the mutual benefit of mankind.

Unfortunately in a world of non-productive top-heavy middle-class administration and financial management it is sometimes difficult to see just how creative in real terms our work actually is. We need only see the white-collar skyscrapers standing where the old blue collar factories once stood to see how much has changed in the last 100 years or so.

When Karl Marx talked so simplistically of the bourgeoisie and the proletariat (lumpen or otherwise) we know what he meant. He also maintained that work was the curse of the drinking classes but this may or may not be true. Yet the old balance between the traditional middle-class and the traditional old working-class has changed beyond recognition. Mainly due to technology the former has increased as the latter has decreased. Most manufacturing is now done elsewhere or sometimes by using relatively cheap immigrant labour where possible.

Whilst we primarily work for the immediate survival of ourselves and families we also work to gain some job satisfaction in collaboration with others for the common good. Work traditionally very much takes the place of hunting, teamwork and male bonding for men in particular. As already discussed the role of women in the workplace has become increasingly important during the past 100 or so years. They have traditionally been paid less than men. These inequalities are now, thankfully, generally less than they used to be but there is still a long way to go in some areas.

Pity everyone is on different salaries which above all is the biggest cause of discontent in the work place. The writer is convinced that amongst the many social upheavals of the 20th Century the so-called emancipation of women must be of major significance to human-kind in general.

True equality for women must soon become reality in most of western society and this is long overdue. It is believed that other social differences between the sexes have become less than they ever were and that this can only improve prospects for survival. On the other hand; many women in eastern cultures are still denied paid work, or the western concept of freedom as we know it and their primary roles as homemakers often remains.

I also believe that the moderating and civilizing influence of women in general has done much to restrain the aggressive macho world of men. Although I believe the macho culture still remains alive and kicking in those parts of the world where the big moustaches still reign supreme.

Analogies between work and hunting are numerous. Sales managers for example may thrive on the chase for that elusive next order. Bankers may make a killing in the stock market. You may see them doing 80 on the motorways pursuing that elusive dollar. Yes, the old cliché of it being a jungle out there keeps on popping up but what would you expect in a treatise like this about survival?

Competition can be fierce and remorseless. This can sort the men from the boys. This is more than just a game. This is serious business. There are many other similarities between the workplace and the jungle for us to think about. Do we even realise working and selling our labour to the highest bidder is the only way to gain any pride in bringing up our families? We want them to do better than we did one day. We have aspirations. Yet they too will sell their labour to the highest bidder.

Many of us in this age of technology (particularly those with manual rather than professional skills) are unfortunately denied the basic human right to creative and well paid occupations altogether. Thus, together with the so-called unemployable of all classes, these marginalised fellow citizens are denied the fundamental right to honestly earn a living and to contribute to the welfare of their families and society as a whole.

This is a high price to pay: where some deemed best qualified and/or the cheapest, in accordance with present day needs, are wanted and those with redundant, expensive or obsolete skills now appear to be largely unwanted. Where are they to go? In spite of skills shortages: training is now pitifully inadequate. It is a bit like that old capitalist marketing ploy of throwing the over-ripe tomatoes away rather than dropping the price.

I am unable to understand all the ramifications of the, patently unfair, huge discrepancies in salaries, income tax and social benefits in general. However I read somewhere that one person may actually be paid hundreds of times more than another and this does not bode well for the overall satisfaction of society at large. What of those who do not even have a very low paid job (below the tax level) never mind a reasonable living? Can one human being really be worth a thousand others?

We all need work to maintain social status. Meaningful work is all about pride and to deprive our fellow humans of this right is not only counter-productive to society in general but an affront to basic human dignity. If I feel strongly about this it is because I believe survival is not just about morality, pride and dignity - it is about feeling we all belong to the same species. I of all people know what a personal struggle I had to establish and develop a meaningful career over many years.

There are simply too many divisions already in society. We have seen all this before during the tumultuous 20th Century and where did that get us? Social injustice, division and poverty have no part in a truly civilised society. Such a society is corrupted by greed and is not conducive to human survival as an inalienable right for all. Survival at its best surely relies on positive attitudes, good morale and the feeling that we can all contribute our fair share to society at large. This is not the much disparaged socialism: it is merely common humanity.

These basic rights are denied to many. This is frankly cruel, divisive and unfair. What was I saying about us humans not always being a very likeable race? The word stigma springs to mind when looking at this sorry state of affairs.

There are variations in the meaning of the word survival depending on the level of subsistence you have to survive on. For a start your long term health and survival prospects may well be diminished by a relatively poor standard of living. Stress levels will certainly be raised by depression and lack of hope for the future. Negative attitudes to life will become inevitable - especially when seeing the relative prosperity of your perhaps better educated, or just more fortunate, fellow humans.

All the negative aspects of social poverty, unrest, crime and violence are simply too big a price to pay for the relative comfort of some at the expense of the others. What to do about it? I just don't know - but we do pay a lot of money to those with the power and intelligence to know just how a better world might be achieved. This goes back to want of care.

Most modern employment uses one machine to do the work of ten or more people: so jobs are not only limited but rely on a more educated workforce than in the past.

On the other hand light repetitive tasks requiring manual dexterity such as electronic component assembly or clothing manufacture have remorselessly devolved on a largely female workforce who have left the land for the burgeoning manufacturing cities of the Far East such as Shanghai. This is the source of much of our cheap labour-saving devices and communications equipment. Labour saving has become labour denying to many.

In addition, supposing they might still have jobs that is, many now survive who would not have survived the hazards of heavy industry. This is due to stringent laws such as the groundbreaking British Health and Safety at Work Act of 1974. Although, how this relates to occupations such as being a soldier in conflict remains to be seen.

However, for both good and ill, the dignity and strength of manual labour now remains a thing largely of a forgotten era. Many sacrificed their health and life expectancy by hard manual labour, shift work and long hours to build the world we know.

Some in the West survive on social benefits, where available, with few opportunities to escape from this humiliating dependence on others who are more fortunate in the lottery of the job-market. Thus education is often seen by many as the best and only way out of this dilemma. However, this opportunity is neither available, nor affordable, nor suitable to many who have already given up all hope of ever finding the type of work to which they might be most suited.

Some wages and working conditions are so pitifully low that only itinerant foreign workers from even more poverty stricken regimes will take them on. This can cause some friction of course but you can hardly blame these immigrant workers for ensuring their own survival and that of their families overseas to whom they must send pitiably small amounts.

You can also hardly blame those workers who feel displaced by cheap labour having worked hard to gain a reasonable standard of living for their families over many years. Again these issues could have been handled in a more sensitive and amicable manner.

The poor and jobless may well live in social housing. This often symbolically has only one way in and out of the estate. Some social housing is very reasonable; other social housing is not quite so reasonable. These are people's homes when all is said and done. Many people on the breadline deserve medals for the way they cope with adversity and are certainly to be admired for the way they cope with life and injustice.

Yes there are scroungers at all levels in society but for most on benefit life is hard. These are the ultimate survivors if we think about it. Do you know what it is like to wear and be glad of the cast off clothes of the rich? Do you realise that your old cooker or TV set has now been put to further service? What is it like to sleep on the floor? When do you have to go to bed to keep warm? There is still a long way to go.

These shortcomings in our society are particularly damaging to the young and this is unforgivable in the opinion of the writer. It is hardly surprising that we now have no-go areas in our cities where poverty, drugs, gangs, crime and other forms of anti-social behaviour have become the norm. Surprisingly, without idealising anything, these marginal societies are not inhabited by another species as our politicians and press would have us believe. They may well be of another genus (homo-unfortunus) difficult for us to identify with; yet certainly of just the same species as ourselves.

This is what I mean about our survival. We must never forget that we must all continue to work, live and survive. Life is too sacred to be so readily squandered for empty so called political principles. This is our birthright: ask any mother. Meanwhile what of our wish to work for our own day to day survival and to feed our families? Socially we become inhibited by our characteristic conservative upbringings, our families and social networks and/or even our regional dialects and are therefore less able to move on and adapt to changing circumstances later in life.

With no house to sell how can you move away for work? Many live literally hand to mouth on a day to day basis; so how can we blame them for lack of foresight in terms of long term survival? It is easy to become hot under the collar about these manifest inequalities but not so easy to know what can be done about it.

Somewhere, someday, regardless of political persuasion, the ongoing insanity of gross divisions and injustice in society may just be consigned to history where it belongs. In terms of a more caring society we may wake up to the fact that wealth and survival prospects will have to be redistributed somewhat in the interests of all concerned.

We cannot remain a society of have-nots and have-yachts. If some of the yachts were slightly smaller and a tad less ostentatious then maybe that would be a start. If some of the have-nots were just a little better off: then that too would be a start in the right direction. If some of the haves were slightly worse off: then that would also be a step in the right direction. You never know it might be a good investment. It would be an investment in survival for all.

By the way if I come over as a much reviled so-called 'socialist' or 'liberal' even; for these quite logical remarks; then think again. What if I were called a humanitarian? Or even a so-called 'Christian': then I would be up there with Jesus! It is simply a matter of caring.

Life can be a challenge for us all. Thus I have much sympathy for those who have their own, often harrowing tales, of survival against all the odds. To say someone is a born-survivor is one of the highest of accolades in my opinion. I would encourage anyone not to ever give up hope for better things, ever.

When people find life to be a struggle for the survival of not just themselves but for their families through no fault of their own we must seriously question why this is. Much of this is the unfortunate downside of our wonderful age of labour-saving technology which undoubtedly benefits some more than others. This is not a Luddite statement simply a fact of modern day life. Others are cynically exploited as so called cheap labour; yet these all too vulnerable people also need to survive.

We don't care enough for our people do we? We must face reality.

Perhaps survival is not quite as fair a concept for all as some of us idealists might like it to be. This is a sad and inescapable conclusion. The efforts of the poor to survive are perhaps more challenging than anything which those of us in our relatively comfortable middle-class lives could even begin to contemplate. Yet we all share common needs.

The poor and underemployed too have aspirations - not least for better education and survival prospects for their children. These are fundamental needs. In fact when looking at human survival needs we sadly fail our fellow human beings at so many levels. Is it so surprising that so many negative attitudes exist.

I, personally, have had about 12 jobs and many different 'homes': so I actually know what I am talking about here. I was adopted at the age of 3. Then my adoptive home became a broken home: not a poor one but a very well-off one. As a child I was up and down so many times that I can only wonder that I never, ever gave in. Stubborn is my middle name and it can only have been character building. My acquired personal and psychological survival values remain with me to this day.

My education was shattered. I ran away to sea. I was poor, lonely and insecure as a young man. I have been homeless and unemployed. I have lived in bug infested accommodation. I have been in poorly paid, unhealthy and repetitious manual jobs. I have been 'skint'. I thought myself a born 'misfit'. Yet I realised that I was so much better off than many. I never starved. I even benefitted immeasurably from it all. If there are gods then mine certainly has a good sense of humour.

My own sense of humour and curious mind have been invaluable. Whilst ever restless: in my twenties, I went to college and then developed a career for myself. This career eventually took me all over the world and gave me a pension too. Above all, I became a happy family man. The care, loyalty and encouragement of my beloved wife and children have been crucial to all that we have achieved together.

My children now have children and the world goes on - surviving. This vividly reaffirms the realities behind my chapters on Caring and Family. I have been greatly helped along the way. My needs have been fulfilled on so many levels that I can only be eternally grateful. My health has held out so far. What more can you ask of life? I progressed from very shaky beginnings to a secure, loving and very fulfilling life

Now past my biblical *'three score years and ten'*: every day is a bonus. I survived - many didn't. Yes, I know about life. Yes, I know about needs. Yes, I know about survival. I was one of the lucky ones.

Earth provides enough to satisfy every man's needs, but not every man's greed.

Mahatma Gandhi

It might be beneficial to look at our human needs as we approach the end of this long and varied programme about human survival. We need to survive and to survive we have needs.

So what does it take to make us satisfied and fulfil our needs? I first looked at the work of Herzberg in terms of human satisfiers and motivators but found this to be more relevant to the work-place than the eternal struggle by the human race to survive on an often hostile planet.

However, Maslow's Hierarchy of Needs, also used by professionals when discussing the workplace, did seem broadly applicable to humanity in general. Well the World is the potential workplace for all of humankind: so why not use it? My point is that we <u>need</u> to survive - simple really. We can therefore summarise our common needs in terms of survival using a simple model. Envisage a neat isosceles triangle with five levels like this:-

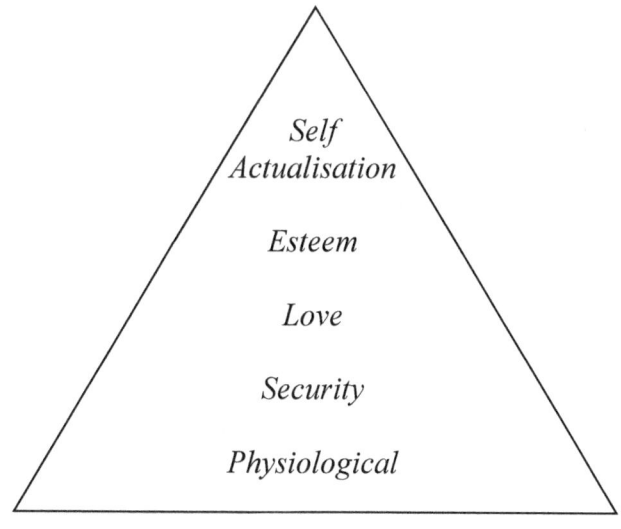

More complex versions of Maslow's triangle show additional sub-divisions in each section. These more detailed facets are beyond the scope of this presentation but we can see that each step level of the supposed pyramid simplistically climbs upward; from basic life survival at the base level; towards more abstruse requirements at the apex.

You will already have deduced how inextricably linked our human needs are to our common welfare and ultimate survival. This almost goes without saying of course. You should immediately perceive many cross references with my ideas about survival in previous chapters.

Firstly, the *Physiological* foundation level, resting on the bedrock of existence, must surely be reflected in my statement that everything we are relates to our survival on earth?

I suppose *Security,* on the second level, is most associated with my chapter about Family. It also incorporates many elements of practical human endeavour in terms of shelter, farming and groupings as stated in my text. These are vital to human evolution and well-being.

The third level: *Love* certainly equates strongly with my essays on Natural Selection, Caring and Family and is crucial to survival.

The fourth level: *Esteem* probably best reflects the importance I place on Social-Status and the previous essay on Work.

Self Actualisation at the very pinnacle may well incorporate peculiarly human metaphysical, intellectual and spiritual notions already discussed at some length. Perhaps this is the *doing* and *thinking* department?

We note that the lower levels relate more directly to practical everyday survival than those at the apex. The lower levels show basic survival needs common to all humans and animals. The higher levels show more esoteric requirements for self-actualization and self-esteem more relevant to the human condition.

However, as in all hierarchies, the pyramid becomes narrower and sharper towards the top. These higher moral and ethical values are over and above the more basic requirements shared by all species. In view of the proven success of humankind to date; it seems that these extra intellectual, aesthetic and moral characteristics must therefore have been of significant value for our survival. This is all very idealistic of course.

I am not sure I agree with all that the triangle suggests. However, this very simplified version of Maslow's Hierarchy of Needs triangle provided some interesting information in terms of survival. Without any denigration of some very noble and intelligent animals it will be seen that humans need some extra impetus for survival.

In fact this triangle does little to address the perennial survival problems of conflict and aggression in terms of territorial maintenance and acquisition. At best it is merely a very basic guide to our very many human needs. However, I believe it to be a very useful tool to illustrate my point that humans require far more than basic animal physiological needs in order to enhance their survival.

The triangle also, perhaps unwittingly and ironically, reflects just how human society is structured - those at the bottom of the symbolic hierarchical triangle merely surviving - those at the top enjoying the finer things in life. Most humans aspire to social status and a reasonable place in the hierarchy - not just for themselves but for their children's, children's children.

Regrettably, some of the very basic humanitarian benefits of the second layer from the bottom are denied to many of our fellow human beings as already suggested. They are struggling to even get their foot on the second level - never mind attain the confidence, esteem and respect of the fourth level. This compromises their hopes for survival and is just not good enough. Remember these are basic human needs. We do need to care if we are all to survive. Survival is not just for the few.

These are fellow human beings not another species. I think they are worthy of better. Nonetheless, this hierarchy of human needs convinces me that survival is more about living than merely existing - and long may it remain so. Survival = Life/Time.

19   CONCLUSIONS

The Seven Deadly Sins of Mahatma Ghandi

Wealth without Work
Pleasure without Conscience
Science without Humanity
Knowledge without Character
Politics without Principle
Commerce without Morality
Worship without Sacrifice

Lastly, I could not resist the wonderful seven deadly sins of Mahatma Ghandi (1869-1948) quoted above. These seem to encapsulate the type of ethical values essential for the spiritual and moral survival of humankind. It is probably not that simple of course.

We can only survive if society is fundamentally good. Yet we do not always treat our fellow humans with all due respect. In essence we are competitive hunter-killers with warrior genes who learned to co-operate and ultimately evolved to grow crops, trade and live in cities.

I believe that mass human conflict is largely due to competition for territory and an over-structured society where our endless pursuit of status makes for too many levels in society and consequent high levels of aggression, inequality, injustice, poverty and resentment.

If we accept Darwin's theories of Natural Selection and Survival of the Fittest and that we live-on in our descendents: then we may also consider if theories of Ancestral Memory are a credible proposition?

Instinct is a powerful contributor to survival. I have decided that human imagination, control, stamina and caring also play key roles in our survival.

Mankind is a virile and flourishing species at the head of the food chain. Tools and animals helped to make us what we are. We do not even think about our day to day survival.

Seemingly limitless natural resources are now under threat. We are victims of our own success. We ignore the land which feeds us at our peril.

Occasionally greed and selfishness do us no favours. Yet we can be charming, ethical, visionary and creative - we are Human. If you have got this far; you will have made your own minds up about human survival by now. If not, then I have failed in my mission.

Finally, before taking the bull by the horns in the Appendix - Defining 'God', what of this somewhat sweeping assumption in the Introduction:-

'Everything we are relates to survival. Everything we do is related to survival. Everything we think relates to survival. It is in Mind, Body and Soul'.

Firstly (and most importantly) everything we are is defined by primary survival through life and time. Whilst taken for granted, this is all about being alive and staying alive through every second of every day. This is also about basic needs such as breathing, eating and sleeping. We are talking of our very existence, Life-force and metabolism. This is about procreation, equilibrium and vitality. Without even thinking about it we continue to sustain and recreate our very beings for our allotted spans on earth. This is pure survival.

In terms of what we do: we can see that secondary survival activities such as work, rest or play are also a routine part of day to day survival. We earn a living, relax and keep fit to play our part in the world around us. We travel, socialise, communicate and see and hear. We balance, develop and exercise our unique upright human physiques. We eat, drink and make merry. We take due precautions to avoid accidents. We use tools to enhance our lives. The automatic body-machine is by turns: idling and burning up energy to make sure we fulfil the basic needs of our species. Every detail of our lives seems to confirm a link (no matter how tenuous) with day to day survival. This truly is survival of the fittest.

When considering everything we think: we can confirm ongoing tertiary survival conscious and subconscious brain activity: such as learning, reading, writing, thinking and dreaming. Then some of us even pray, fantasise or daydream. We try (in vain) to work out some answers to the Great Questions. We laugh, cry and have emotions and feelings. These are all vital for psychological well-being. Then what of our ids, egos and super-egos: the powerhouses of our minds? In other words we never switch-off day or night as any writer will confirm.

Mind, body and soul never take a break. These are the Holy Trinity of survival and long may they remain so. Thus intellectual, physical and emotional survival is assured. Therefore my major conclusions about human survival are as follows:-

I.   *Everything we <u>are</u> relates to survival.*

II.   *Everything we <u>do</u> is related to survival.*

III.   *Everything we <u>think</u> relates to survival.*

And finally:-

IV.   *Good <u>must</u> outweigh evil for universal human survival.*

APPENDIX - DEFINING 'GOD'

The energy of the mind is the essence of life.   *Aristotle*

DEFINING 'GOD' - INTRODUCTION

Let us begin by asking if 'God' is an ill-defined and misconstrued word? Is God a vague human hypothesis for that which is essentially unproven? Is God a state of mind - an idea? The word is ambiguous and implies something as mystical and ethereal as to be often called the Great Unknown.

Yet the word God still means so much for so many. God may even be construed as all things to all people. Some even make a career from these beliefs. Several people give their lives to their vision of God. They have an instinctive compulsion to worship this most indefinable concept.

God is literally and metaphorically associated with Good. Most cultures interpret their own version of God or Good a bit differently from each other. Who is right? Confusingly the perception varies from the universe to the soul.

Commonly Gods are seen as somehow human or super-human. To others there are a profusion of spiritual Gods. This is all too perplexing and presents us with an almost impossible dilemma. In western culture there seems to be two main perceptions of God as follows:-

1. *The Inner divinity of the human soul, or mind.*

2. *The Outer universal Creator of matter.*

Both of these disparate key versions interpret the idea of the Life-force or Being in a very different way. The word Holy is often used to give an extra spiritual or ethereal dimension to these concepts: thus encompassing many differing elements above and beyond basic human language or understanding.

Thus superstitious medieval perceptions of God; which still exist in our modern scientific world, can be mystifying to say the least. Yet the seed remains. Again, on the one hand; there is the inner concept of man in the guise of a human being. On the other hand; there is the outer concept of nature or being. However, they both relate strongly to the mysterious Life-force at the heart of human survival. This makes defining this ambiguous word quite a challenge.

So let us compromise between science and religion by looking for some common ground between these disparate notions. For the purposes of our theories about human survival let us assume God is really some kind of universal Energy derived from matter of which the human life-form is merely a fragment. Yet, if the human persona is the one which is uniquely capable of imagining these things; it once again becomes a question of mind over matter*.

When discussing matter attention is drawn to the Periodic Table of the Elements where each known chemical element is listed according to proton derived atomic numbers, electron configurations and recurring chemical properties. It is the Genesis of all matter.

DEFINING 'GOD' - DISCUSSION

Let us presume that a hypothetical great and unknown Creator equates to Energy, which is present in both heaven and earth. Where does that take us? There is confusion between the universal energy out there (heaven) and our own down to earth internal human Life-force (earth). It is almost as if there is a superhuman Father (God) out there and a more human Son (Jesus) down here.

Yet, are we just guessing because we don't really know? Humanising it all? We do this don't we?

How can we associate the human God, or Life-force with that small part of the vast universal energy of which life on earth is only a miniscule representation?

Remember that humans only form a diminutive part of the overall scheme of things on earth. Bear in mind the 8,750,000 other living and surviving species. Remembering that we share life with so many other animal and vegetable species - what makes us have a special direct line to a supposed survival deity and not them? We are all fellow survivors after all. Is the supposed deity only for the most select predatory survivors?

What is life? Was life created by design or by evolution? Is it part of a great plan or pure chance? Apart from the dreadful praying mantis do animals ever pray? Do the horrific preying crocodiles have a God or do they epitomise the devil? Or are they simply and cold-bloodedly surviving without quite knowing why? Do they survive as part of nature or because they are bigger, fiercer or even perhaps 'holier than thou'? Do the hunters have a deity denied to the hunted? Win or lose?

Or is the concept of God a human control device for something we can only guess at as being human or superhuman without any real proof? Remember our propensity to control? Or is it just all in our inflated egos and hyperactive imaginations to think that we actually know? How honest are we? Are we really so arrogant as to think we know - I mean really truly know? Yet we survive and survive and survive.

All life forms seem to survive regardless of species whether they have an imagined God or not. If you persuade me that Nature is another name for God then I would be more convinced. Nature is pure living energy if you think about it. Do we need our gods simply because we can imagine a better world - pure psychology - or pure survival in other words? Yet energy by whatever name you may call it does exist and so do we.

We live and survive and we call it the mysterious Life-force. Survival and Life-force are inextricably linked. The Life-force, critical to the survival of humans in particular, has featured throughout this study. If for the sake of argument the human vision of God, in the widest meaning of the word, is bound up with all the great energy of the universe then what of the everyday inner man/woman i.e. the Life-force?

Because the writer, who is a typical human being, does not know what or who it is that a devout worshipper actually sees, hears or feels during prayer; then how can we truly identify the image other people have of their chosen God? Do some have special insight denied to the rest of us lesser mortals - Predestination perhaps? Are they the destined, chosen and special ones? Is God so unfair?

What makes God real to some; yet not all of us? Why do we have so many different ones depending on where we might live? Do we really know? I mean know? Do we see the Christian conception of God as a father through the image of a man, say Jesus - or the image of a woman, say Mary - as son and mother of God respectively? Or do we see a ghostly entity in our imaginations; which is not a person but something otherworldly as stimulated by our fellow worshippers and/or by symbolic images such as holy relics, crucifixes or icons?

Or is it pure fear? Something metaphorically generated deep in the mind - something mystical, spiritual and otherworldly - rather than something that is more logically explainable in terms of the physics of the Life-force?

As already discussed; the origins of all matter, energy and the consequent evolution of the Life-force on earth, remain as intangible, inexplicable and mysterious as ever. No matter how much natural intelligence and imagination we humans bring to this great question we may well have to resort to some remnant mystical notions of a supposed Great Creator, or the unsolved scientific idea of being for now.

On the other hand, if we can accept Einstein's theory of Energy being neither created nor destroyed $E = mc^2$ (once matter exists in any case - that is) then we can see that the Life-force energy will remorselessly continue in one way or another down the generations. This is the essence of survival.

Assuming we wish to believe in a mystical entity, deity, controller or creator, in lieu of any other explanation; then how is this manifest in our day to day lives? How do we reconcile our all too human doubts, fears and uncertainties in our minds? How do our egos, super-egos, ids, souls and personalities in general, deal with these intangible conscious and subconscious imaginings?

What kind of God are we looking for in fact? And where is he/she? Is God a mere cop-out - an admission that we don't really know - or a living tangible being? How is she manifest to those who can find no other explanation? Where is he? Do we perceive ancient dream-like mystic apparitions from deep in our ancestral memories in fact? If so, why have so many of us been excluded from these hypothetical supernatural voices in our own imaginations?

We poor doubters too are sincere seekers after truth: so why do we not all hear these voices if we are all truly one under God? Are perceptions of an ethereal God actually figments of our all too powerful imaginations? Thus convincing ourselves that we are in the know and special in some way? We all need to feel special - it's part of our all important need for social status of course.

Is that what religion is all about - feeling special? By feeling special we certainly feel better and can see a better world out there. It can only do us good and that is what life needs. It needs to be as good as we can make it to ensure our survival. It is pure psychology for optimum, or optimistic, survival once again.

Is there an inescapable problem impeding human understanding? We can only envisage the vast universe from a limited and very subjective human perspective. Are we blighted by our own humanness? Human language does not even begin to express the truths by which we are confronted.

As soon as we hear emotive words such as God, power, life, etc, we instantly put inadequate human interpretation on these notions. All the centuries of myths and legends swamp our brains from a realistic vision of what we are trying to envisage.

We simply cannot stop seeing things from a purely human and egocentric view point. We feel empowered. We really see God as a bigger and better upright fellow human being - don't we? We can't stop ourselves. We are surrounded by things which we can only guess at. Our language is not yet adequate to express all that remains to be discovered - for which no modern words can do justice. So in the light of these misgivings I can only draw your attention to the essential roles of human intelligence and imagination.

My own theory concerning the psychological importance of human intelligence with regard to the Life-force is as follows.

This is more difficult for me to communicate than it is to comprehend so please bear with me. We will have to stretch our imaginations to the very limit in order to pursue this ambiguous concept.

I said earlier that it was difficult to imagine exactly what it might be that a religious person was seeing, hearing or perceiving when they prayed to their perception of God. In other words how much of it is in the imagination?

What if, for the sake of argument, we envisage all that presumed cosmic energy/spirit out there to be one huge mystical kind of Intelligence? Our highly developed subconscious human imaginations may even perceive this presumed 'Universal Intelligence' as not only logical but worthy of worship? We impose our human brain power on everything else. It is all in our egos. In other words we worship ourselves. We worship our own intellects and all that they may imagine. Is it all in our imaginations?

Back to the so called figments - yet all very real and vivid when combined with our strong emotions - all in the mind again? Our hugely overactive human imaginations/egos probably induce us to see everything in human terms. Therefore it would be quite in character for us to assume that any supposed human deity has human or superhuman qualities.

We would assume this supposed super-Intelligence, or super-Imagination if you prefer; to be capable of creating and controlling life on earth. Therefore we can readily appreciate that our own intelligence/imaginations which are contained in each and every one of us are of godlike significance. This brings us back to it all being in the mind.

If we follow the logic that we may in fact be unconsciously worshipping the Life-Force/Imagination itself (by another name); as distorted or interpreted by centuries of subconscious ancestral memory and folklore, when we knew nothing about energy: then perhaps we are not far wrong in our assumptions? This holds particularly true to this day when we still do not know all there is to know about the mysterious Life-force. Yet we have come a long way from our old ignorance, superstition and supposition.

We live in a world dominated by the laws of science, chemistry, engineering, electronics and mathematics. We fly, we communicate across the world in seconds and we have computers for starters. We have discovered how to utilize much of what the world has to offer in the way of metals, fossil fuels (ex-life) and other present day life forms. Well some of us perhaps, now understand a bit about the energy all around us even if we do not know its purpose.

So what if we look at the nature of the Life-force in the light of present day knowledge?

Well we can safely attribute the Life-force to astral power as represented by the heat and electro-magnetic energy of the sun and the boundless hidden resources of the earth. We can immediately see that physics, chemistry and biology are the Holy Trinity of being. If we sit and think about it we can see that all these were unknown to our forebears who could only worship, wonder and pray to these great unknown and magical powers.

Now we can see that, although science cannot explain everything: there are some infallible laws of science and being as already discussed, such as Einstein's Laws of the Conservation of Energy: ($E = mc^2$). From this we can deduce that fundamental universal energy is readily absorbed by all living beings. We can further confirm that our human body energy systems are a mixture of biological, chemical and electro-magnetic forces. We have even invented organic medicines, chemicals and computers to enhance our survival.

The make-up of our amazing human bodies reflects all these known forms of energy. Our physical being is structured to enable survival peculiar to the planet we live on in common with all our fellow travellers through space and time. The main ingredients for life are thought to be oxygen, hydrogen, nitrogen and carbon as reflected in the basic analysis of our bodies discussed in an earlier chapter.

The human Life-force is what keeps us alive. It has developed over countless generations and is truly remarkable, miraculous and phenomenal. The Life-force is very many things. Much is already understood, some may be even further researched and some aspects may never quite be fully comprehended. All we can say is that life is of an incomparable value which must be continued for as long as blood pulses through our veins.

As humans our pumping hearts transmit life sustaining chemical blood to our muscles, limbs and bones for survival. Our breathing feeds our bodies with life sustaining oxygen for survival. Our mouths and stomachs process the organic food and drink essential to give us living energy for further survival. Our wonderful and phenomenal brains above all give us logical, intellectual and spiritual energy to make our survival so spectacularly successful.

Our hearing, seeing, smelling and touching senses are incomparable tools for survival. Computers may owe their origins to the vision of our super brains and conversely our brains are truly like computers.

The human personality; as repository of the id, ego, super-ego, God and memory, lies firmly in charge of all that makes us such an intuitive, imaginative, cognitive and visionary species. The heart, brains and nervous systems together give us powerful and amazing internal and external electro-magnetic impulses. Our reproductive systems further ensure the continuance of our unique species now and long into the future. We nurture, love and care for our children.

We may have many faults, weaknesses and imperfections but so long as we retain the moral and ethical capacity to care for each other then we can be proud to be humans no matter how we choose to express our gratitude to whatever kind of energy, spirit or deity may be out there ensuring our survival - day after day, surviving to live and living to survive.

The world's great religions are ancestral spiritual control systems at the heart of civilisation. Therefore, it is hard to deny the wisdom of centuries or the faith of millions. In lieu of science many fervently worshipped the great Unknown as a perceived mystical entity universally known as God. To Christians the Son of God is known as Man or Messiah. I can't accept that God is man but I recognize that God might be in man as living energy.

Some say religion is all in the mind. Or is it a question of mind over matter? God is perceived to be the Creator yet Einstein stated that matter can neither be created nor destroyed. However, matter can be transformed into kinetic energy ($E = mc^2$). Thus we can feasibly associate God with residual cosmic energy derived from matter. So, in lieu of other explanations; traditional and instinctive worship of the heat, energy, time and light of the sun, moon and stars became a logical thing to do. Therefore, to think of God as energy, or spirit, or all things if you prefer, is a realistic and attractive proposition.

Without energy (spirit) there would be no being - without being there would be no life. On the other hand why matter and energy exist is infinitely beyond any human understanding. If we consider that God only lies dormant in matter (mass) then we see Him as being merely of unfulfilled potential energy without which organic/genetic life could not exist.

If, on the other hand; we see Him as of derived kinetic energy, through space and time: then we see Him as a dynamic continuing entity as manifest by evolution of life on earth.

This plausible idea may be acceptable to both religion and science summarised as follows:-

If God = Dormant matter (mass) then He is only of latent <u>potential</u> energy

If God = Volatile matter (or kinetic energy) then He is <u>living</u> energy

If God = Living energy (through space and time) then He is <u>dynamic</u> energy

Therefore science pragmatically equates to the human concept of a presumed God in terms of matter being converted into dynamic energy through space and time as follows:-

GOD = ENERGY / SPACE and TIME   or $E = mc^2$

If God is energy and/or in the mind too, then he is virtually all things to all people - thus neatly encompassing both the power of universal being on the one hand and the power of the human mind or imagination on the other. The human mind is capable of imagining almost all things but not quite as we know only too well. Some things we will never know.

So my tentative but hopefully well argued metaphysical conjectures are speculative to say the least. I suppose everyone has their own beliefs and who am I to know? I mean REALLY KNOW. Maybe I am just guessing - but no more than anyone else I suppose?

DEFINING 'GOD' - SUMMARY

- *We can credibly define the human concept of God as living energy or 'Being'.*

- *As human beings we can envisage the universe as the source of all matter.*

- *Matter is the basis of life (which is possibly a form of motion or 'kinetic' energy?).*

- *Energy can neither be created nor destroyed ($E = mc^2$) so matter is eternal.*

- *What of matter? What of eternity? What of time? What of space? What of being?*

- *Even our supercharged human imaginations can surely never understand all of this?*

- *So we created the concept of a human-like God to explain it all - mind over matter.*

- *Thus, by having a God we feel better about ourselves in order to survive.*

What we do <u>not</u> know, of course, is WHY?

Epilogue

An aura of adrenaline fired common, religious and emotional human Energy is radiated during communal worship. This may be thermo-bio-magnetic or telepathic by nature - yet is powerful and palpable. As with other species the mass instinct of the majority then takes over from the individual. So the accumulated energy, or Spirit of a perceived common God is generated by those who truly believe and have unreserved faith in their chosen deity.

or

Others more pragmatically worship at the secular feet of the great philosophers and scientists: Aristotle, Darwin or Einstein. We are all in the world together. We live. At worst we let ourselves down - badly. At best we all have unique human egos, imaginations, curiosity, and humanity. Thus does Humanity Survive.

'Cogito ergo sum'
'I think therefore I am' – Descartes

www.ingramcontent.com/pod-product-compliance
Lightning Source LLC
Chambersburg PA
CBHW051912170526
45168CB00001B/348